THE GREENHOUSE EFFECT

THE GREENHOUSE EFFECT

HAROLD W. BERNARD, JR.

BALLINGER PUBLISHING COMPANY
Cambridge, Massachusetts
A Subsidiary of Harper & Row, Publishers, Inc.

QC 912.3 .B47

Bernard, Harold W.

The greenhouse effect

The opinions, findings, and conclusions in this book are those of the author alone (except as noted) and do not necessarily reflect those of any of the author's past or present affiliations.

International Standard Book Number: 0−88410−633−0

Library of Congress Catalog Card Number: 80−13125

Printed in the United States of America

Library of Congress Cataloging in Publication Data

Bernard, Harold W
  The greenhouse effect.

  Bibliography: p.
  Includes index.
  1. Greenhouse effect, Atmospheric − United States.
2. Climatic changes − United States. I. Title.
QC912.3.B47   551.6   80−13125
ISBN 0−88410−633−0

DEDICATION

To my brothers, Rod and Rick,
who have never had to deal
with environmental adversity,
but who know a lot about
handling personal adversity.

CONTENTS

List of Figures   xiii

List of Tables   xv

Acknowledgments   xvii

Introduction   1

Chapter 1
The Greenhouse Threat   5

Increasing Amounts   5
A Super-Interglacial   7
A Cooling Cycle   8
The Impact on Our Lives   11

Chapter 2
We Can't Put the Weather in a Test Tube   13

Historical Trends   13
Future Trends   16
Nervous Breakdowns and Other Problems   17

Feedback   19
How Much Proof?   20
How Fast?   20
Other Theories   22
The Ice Age Cometh?   23

Chapter 3
A Search for a Climatic Analog   27

The Westerlies   27
Dust Bowls and Ice Packs   32

Chapter 4
The Return of the Dust Bowl   35

Droughts and Sunspots   35
Black Blizzards   37
Snapshots   38
Eighty-Five Bushels Per Acre   41
Wheat, Rust, and Prickly Pears   43
The Ogallala Aquifer   44
The Resilient Crop   45
The Tillamook Burn   45
Will We Have Enough Water?   46
Synfuels and Catch-22   48
Sixty Cloudless Days   50

Chapter 5
The Long Island Express   51

Hurricanes   51
Tornadoes   54

Chapter 6
Climatic Stress   57

Heat Waves of the 1930s   59
The Dangers of Heat   63
Forty-One Days Below Zero   64

Drownings in the Dust Bowl   67
Now is the Time   69

Chapter 7
The West Coast and Alaska
—the 1930s   71

The West Coast   73
Alaska   77

Chapter 8
The Intermountain West, the Rockies,
and the Southwest—the 1930s   79

The Intermountain West   79
The Rockies   82
The Southwest   83

Chapter 9
The Great Plains and Texas—the 1930s   87

The Northern Plains   87
The Southern Plains and Missouri   94
Texas   96

Chapter 10
The Great Lakes, the Ohio Valley,
and the Lower Mississippi Valley—
the 1930s   101

The Great Lakes and the Ohio Valley   101
The Lower Mississippi Valley   106

Chapter 11
The Eastern United States—the1930s   109

The Northeast   109
The Middle Atlantic States   114

The Southeast   116
A Review: The Dust Bowl Decade   119
A Hemispheric Perspective   124

Chapter 12
The Ultimate Pollutant   127

English Wines   127
A 200-Year Drought   129
The West Antarctic Ice Sheet   130
Alaskan Camels   131
The Age of Reptiles   133
Crocodiles in New York   135
Fifty Years to Switch Fuels?   136
Winners and Losers   137

Chapter 13
The Energy Outlook   141

Growing More Trees   141
Our Energy Future   143
Oil   144
Natural Gas   146
Coal   147
Nuclear   149
Solar   150
Biomass   152
Hydro   152
Wind   153
Ocean   154
Geothermal   154
Hydrogen   155
The Best Resource   156

Chapter 14
What Can We Do?   159

Action Before Knowledge   159
An Amalgam of Issues   161

The Unsophisticated Technology   162
"A Damned Good Decision"   163
Tax Credits and Golf Carts   164
Gas and Nukes   166
Doing Better: National Goals   167

Notes   171

Index   183

About the Author   189

LIST OF FIGURES

1-1   A schematic depiction of Northern Hemispheric
       temperatures as suggested by studies of Greenland
       ice core samples   9

1-2   A schematic depiction of the 180-year cooling cycle
       since 1800   10

2-1   The atmospheric CO_2 content as measured at Mauna
       Loa Observatory, Hawaii, between 1958 and 1977   14

2-2   Calculated atmospheric CO_2 concentrations from the
       beginning of the industrial revolution to the year 2000 15

2-3   The projected global temperature trend   21

3-1   The undulating pattern of the westerlies   29

3-2   The jet stream during the Medieval Warm Period and
       the Little Ice Age   30

3-3   Northern Hemisphere temperature trends since 1880   32

4-1   Annual smoothed sunspot numbers with the peak years
       of major western droughts   36

4-2   A "snapshot" of the 1930s drought at its peak in
       July 1934   38

4-3   A "snapshot" of the 1950s drought near its peak in
       July 1956   39

4-4   A "snapshot" of the 1970s drought in April 1977   40

4-5   A plot of simulated corn yields using 1973 technology
       and harvested acreage   42

5-1   The annual variation in the number of tropical storms
       and hurricanes in the Atlantic, Caribbean, and Gulf of
       Mexico, 1885–1978   52

6-1   The average temperature deviations over the United
       States during the record hot month of July 1936   61

6-2   The average temperature deviations over the United
       States during a modern hot month, July 1974   62

7-1   The location of the cities whose weather records were
       studied as the basis for a comparison of the climatic
       averages of the 1930s with current normals   72

9-1   The mean July temperature differences between the
       1930s and now   88

11-1   The pattern of average yearly precipitation in the
       1930s as a function (percentage) of current normals   120

11-2   The extent of the warming (and cooling) of the 1930s
       relative to modern normals   121

12-1   An estimate of the precipitation regimes that accompa-
       nied the Post-Glacial Optimum   132

12-2   The consequences of a 250-foot rise in sea level for the
       eastern United States   136

LIST OF TABLES

6-1		58
7-1	Monthly Averages, Salem, Oregon	74
7-2	Monthly Averages, San Francisco, California	75
7-3	Monthly Averages, Los Angeles, California	76
7-4	Monthly Averages, Fairbanks, Alaska	78
8-1	Monthly Averages, Walla Walla, Washington	80
8-2	Monthly Averages, Salt Lake City, Utah	81
8-3	Monthly Averages, Winnemucca, Nevada	82
8-4	Monthly Averages, Sheridan, Wyoming	83
8-5	Monthly Averages, Denver, Colorado	84
8-6	Monthly Averages, Phoenix, Arizona	85
8-7	Monthly Averages, Albuquerque, New Mexico	86
9-1	Monthly Averages, Havre, Montana	89
9-2	Monthly Averages, Bismarck, North Dakota	90
9-3	Monthly Averages, Minneapolis–St. Paul, Minnesota	91
9-4	Monthly Averages, Valentine, Nebraska	92
9-5	Monthly Averages, Lincoln, Nebraska	93

9-6   Monthly Averages, Dodge City, Kansas   94

9-7   Monthly Averages, Oklahoma City, Oklahoma   95

9-8   Monthly Averages, Columbia, Missouri   96

9-9   Monthly Averages, Midland-Odessa, Texas   97

9-10   Monthly Averages, Waco, Texas   98

9-11   Monthly Averages, Galveston, Texas   99

10-1   Monthly Averages, Sault Sainte Marie, Michigan   103

10-2   Monthly Averages, Detroit, Michigan   104

10-3   Monthly Averages, Rockford, Illinois   105

10-4   Monthly Averages, Cincinnati, Ohio   106

10-5   Monthly Averages, Memphis, Tennessee   107

10-6   Monthly Averages, Baton Rouge, Louisiana   108

11-1   Monthly Averages, Burlington, Vermont   110

11-2   Monthly Averages, Rochester, New York   111

11-3   Monthly Averages, Milton (Boston), Massachusetts   112

11-4   Monthly Averages, New York, New York   113

11-5   Monthly Averages, Washington, D.C.   114

11-6   Monthly Averages, Greensboro, North Carolina   115

11-7   Monthly Averages, Rome, Georgia   117

11-8   Monthly Averages, Apalachicola, Florida   118

11-9   Monthly Averages, Lakeland, Florida   119

13-1   U.S. Energy Sources—1977   143

13-2   U.S. Energy Sources—2000   144

14-1   U.S. Energy Sources—2000   167

14-2   U.S. Energy Sources—2000   168

ACKNOWLEDGMENTS

A number of important people contributed their time and knowledge to *The Greenhouse Effect*. Dr. W. Lawrence Gates, director of the Climatic Research Institute at Oregon State University, Corvallis, and Kerry H. Cook, a graduate student in the Department of Atmospheric Science at Oregon State, patiently answered my questions and critiqued key chapters of the book.

Dr. Kirby J. Hanson, director of Geophysical Monitoring for Climatic Change, Environmental Research Laboratories, Boulder, Colorado, also responded willingly to a number of questions.

Dr. James S. McNiel, Jr., president of Mobil Tyco Solar Energy Corporation, Waltham, Massachusetts, and A. Spencer Taylor, manager of Crystal Growth Operations at Mobil Tyco, spent considerable time with me discussing solar energy and energy in general; they also reviewed the energy-related chapters in the book.

David Rockford, nuclear security coordinator at Stone and Webster's Cherry Hills (New Jersey) Operations Center, reviewed my writings on nuclear power and volunteered several helpful comments.

David B. Spiegler, a certified consulting meteorologist in Lexington, Massachusetts, took time out of his busy schedule to critique various chapters at my request. Robert Lautzenheiser, New England Climatic Service, Reading, Massachusetts also helped.

At Environmental Research & Technology, Inc. (ERT), Concord, Massachusetts, the following people aided my efforts: Dr. Brian Murphy, Ernie Abrams, Ed Hummel, Elizabeth Harrington-Bender, Brenda Allen, and Deanna Lynn. My thanks also to the Acton (Massachusetts) Public Library.

Susanah H. Michener (ERT), back by popular demand, produced the graphics for *The Greenhouse Effect*; and Thelma and David Westlake did the typing.

Food services were provided by my wife, Christina, who claims she had to slip dinners under the door to my study while I worked on the book. It isn't true. I always came out for food. But I did welcome the numerous cups of hot coffee Chris delivered, though not under the door.

INTRODUCTION

Before the end of the next four decades, easily within the lifespans of many of us, the climate of the earth may be warmer than at any time in the past thousand years. By the middle of the next century, within the lifetimes of our children—or certainly of their children—it is possible that the earth will be warmer than at any time in the past 125,000 years.

This change, in terms of a geological clock, would happen with lightning quickness. In the course of a few decades the world's climate would undergo alterations that previously took place within a time frame measured in centuries. To say that human adjustment to these changes would be difficult is at best an understatement.

This ominous prophecy is not the misty vision of a doomsday seer or of some headline-grabbing maverick meteorologist. It is the preliminary view of recognized scientists who have studied the effects of the world's continued reliance on fossil fuels—coal, oil, and natural gas—as its primary energy source. The prognostication is based on the best scientific knowledge and thinking available today. In a word, that knowledge tells us that the carbon dioxide released from the burning of fossil fuels may accumulate in the atmosphere to the point where it begins acting like a greenhouse, allowing the sun's energy in, but not letting the earth's heat out.

If the scientists are right—and there is little current evidence that they are wrong—then the world is faced with some critical and immensely difficult decisions over the next few years. Let me interject here that, in my usage, the "world" is not an abstraction. The "world" isn't "the other guy." It isn't some distant, supposedly omnipotent, government bureaucracy. And it isn't a group of dedicated researchers paying homage to banks of number-crunching computers. The world is you and me and our next-door-neighbors. We are part of the problem and can be part of the solution. We can have a voice in the decisions that must be made.

The decisions, basically, will determine our energy future. There is no question that we must wean ourselves from fossil fuels. Climatic warming aside, the fossils are expiring resources. We have already felt the economic and social pressures of limited oil reserves. (The reserves may be large, but they are still limited, and most of them are not ours.) We have felt the first panic as lines formed at gas pumps, service stations closed, and gasoline prices gnawed at our wallets. The comfortable, energy-plentiful environment is vanishing. Unless we make the right choices, things will get considerably worse.

With what do we replace coal, oil, and natural gas? Nuclear energy? The emotional cry is "no!" Solar power? There is no concentrated push in that direction now, and many experts do not expect solar power to be an important energy contributor even by the end of this century. Stringent conservation? Not if it means surrendering our cars and our air conditioners.

The situation is not good. But it is far from hopeless. We have the opportunity to take action that *can* mitigate the greenhouse effect, and at the same time reduce the threat of petroleum blackmail.

This book details the possible climatic consequences of the continued dominant position of fossil fuels in world energy production. It also discusses actions that we can take. The consequences may not be forestalled without sacrifice. In a sense we may be in combat, and the enemy will be us,[1] of course—perhaps the most stubborn foe of all.

But if we understand the cost of losing the battle, the campaign may be carried forth more vigorously. That is the prime reason for this book. It describes, on a region-by-region basis, the specific fossil fuel- (carbon dioxide-) induced climatic changes possible in the United States within the next few decades. The outlook, particularly for the Midwest, is grim.

At best, we can probably delay the effects. It is not likely that we can prevent them entirely. At least by delaying them we can buy time in which to attempt to adjust to them. For ultimately, if it is left unchecked, the human-caused climatic warming could be disastrous. But the ultimate changes probably can—if we act promptly and responsibly—be averted.

Finally, I want to emphasize that the topic of *The Greenhouse Effect* is of central concern. The British scientific journal *Nature* in its May 3, 1979, issue put the subject into harsh perspective: "The release of carbon dioxide to the atmosphere by the burning of fossil fuels is, conceivably, the most important environmental issue in the world today."

1 THE GREENHOUSE THREAT

Carbon dioxide (CO_2) is not normally considered an atmospheric pollutant. We cannot see CO_2, we cannot smell it, it does not damage our respiratory system, and it does not change the color of the sky.

Yet, because of man's reliance on fossil fuels—oil, coal, and natural gas—for energy, CO_2 may be far more dangerous than any of the gases or solids previously identified as atmospheric pollutants.

Carbon dioxide is not in itself harmful. It is found naturally in the atmosphere. By volume it comprises about 0.03 percent of the air we breathe. The two largest constituents of the atmosphere are nitrogen and oxygen. Nitrogen accounts for about 78 percent of the atmosphere's composition, oxygen for just a little under 21 percent.

Carbon dioxide is what we exhale when we breathe. Plants and trees use CO_2 in the photosynthesis process. That is the process that combines CO_2 and water in the presence of chlorophyll and sunlight in order to manufacture carbohydrates. All life depends upon photosynthesis, either directly or indirectly. That is because all animals feed either on plants (carbohydrates) or on animals that eat plants. Obviously we need CO_2.

INCREASING AMOUNTS

The problem is, there is a lot more of it around now, and that—scientists have recently come to recognize—we do not need. There is

5

a lot more of it now because our activities, specifically the combustion of fossil fuels, are injecting more and more CO_2 into the air every year.

Fossil fuels are products of the fossilized remains of plants and trees. When we burn coal and oil, the CO_2 that was absorbed by plant life eons ago is released back into the atmosphere. But it is released at a much faster rate than plant life can use it. The result is that the concentration of atmospheric CO_2 annually increases by a few tenths of one percent. Altogether, the air now holds about 15 percent more CO_2 than it did a hundred years ago.

√Because the world's population is growing—currently at the rate of 80 to 90 million people per year, or about four times the population of California—the demand for energy burgeons at an ever accelerating rate. More coal, oil, and natural gas are burned each year in order to satisfy that demand, and increasing amounts of CO_2 are forced into the atmosphere. Scientists would say this increase is *exponential* and not linear. That is, the concentration of airborne CO_2 does not increase by the same amount each year, it increases by a larger amount each year! A recent study by the National Academy of Sciences (NAS) found that by the end of the twenty-second century, the atmospheric CO_2 concentration might be four to eight times what it is now. [1]

But if carbon dioxide is not normally considered an atmospheric pollutant, and is not in itself harmful, why may increasing amounts of it be dangerous? The answer lies in a physical property of CO_2; it is relatively transparent to solar radiation—sunshine—but relatively opaque to the earth's heat radiation. Or, saying it another way, CO_2 allows sunshine to heat the earth but then traps much of the heat near the earth's surface, rather than permitting it to radiate back to space. This *greenhouse effect* warms the earth. This is not a problem, of course, as long as the amount of CO_2 in the air remains fairly constant; the amounts of incoming sunshine and outgoing heat remain in balance, and our climate remains relatively comfortable.

However, the amount of atmospheric CO_2 is increasing, and most scientists fear that this may lead to a significant warming of the earth's climate. Current estimates suggest that CO_2-induced warming may account for about a 1.8°F rise in global temperature by early next century. [2] Within a hundred years, global warming could be on the order of 11°F, with temperature increases in polar regions as much as three times that. [3]

A SUPER-INTERGLACIAL

Dr. Wallace Broecker of the Lamont-Doherty Geological Observatory at Columbia University, one of the foremost authorities on the carbon dioxide issue, perhaps describes it better when he says the earth would be plunged into a "super-interglacial," with temperatures warmer than anything experienced in the last million years.[4] Sea levels would rise and agricultural belts would be shifted by changing weather patterns. The NAS notes that "for some countries with marginal agriculture, the impact on food production could be severe."[5] Dr. Robert M. White, head of the Climate Research Board of the U.S. National Research Council, puts it this way: "There would be winners, and there would be losers. A climate change could be the cause of a major redistribution of wealth, and from the point of view of mankind, quite an arbitrary one." White, speaking at the World Climate Conference in Geneva, Switzerland, in February 1979, didn't say it directly, but he seemed to be warning that a significant shift in weather patterns—resulting from a warming earth, for instance—could lead to serious international tensions.[6]

If significant warmth resulting from the CO_2 greenhouse effect were to persist over a number of centuries, the Greenland and Antarctic ice caps would melt, and oceans would rise, flooding the world's coastal cities. A recent popular novel fictionalizes the CO_2 problem in a much accelerated time frame in which the message is clear: warmer is not necessarily better.[7]

Some scientists are even blunter about the consequences. Dr. W. Lawrence Gates, director of the Climatic Research Institute in Corvallis, Oregon, warns that if CO_2-loading of the atmosphere continues at its present rate, global warming "may amount to a climatic catastrophe" in the twenty-first century.[8] Other researchers are sounding a call-to-arms. Dr. Gordon J. MacDonald, professor of environmental studies at Dartmouth College in Hanover, New Hampshire, and chairman of a committee that prepared a study called "The Long-term Impact of Atmospheric Carbon Dioxide on Climate" for The U.S. Department of Energy, states, "The Government must start dealing with this problem now."[9]

David H. Slade, manager of the Carbon Dioxide and Climate Research Office at the Department of Energy, still has some doubts about the consequences of increased atmospheric CO_2, but acknowledges, "Everybody agrees that the potential for a serious problem is there."[10]

The problem was made even more serious in mid-1979 by President Carter's decision to launch a major national program to stimulate the production of synthetic fuels. Synthetic fuels—primarily oil and gas obtained from coal, and oil squeezed out of shale rock—are even "dirtier," in terms of CO_2, than natural fuels. In a report to the Council on Environmental Quality in July 1979, a group of scientists warned that CO_2 emissions from synthetics made from coal would be almost 50 percent higher than those resulting from burning coal directly, and almost double those released when an equivalent amount of energy is taken from natural gas.[11]

There is little doubt now that the greenhouse effect has become the greenhouse threat. Unfortunately, modern life seems to be characterized by threats of disaster, and we tend to ignore headlines such as "Climatologists Are Warned North Pole Might Melt" (*New York Times,* February 14, 1979) as just more fodder for Hollywood science-fiction. Indeed, none of *us* may live to see the north pole melt, but our grandchildren might.

Peter Stoler, discussing technical writing in *Time* magazine once said, "Like squids, scientists protect themselves with clouds of impenetrable ink." But on the CO_2-climate issue the ink seems to have cleared. Scientists are speaking clearly to us, and in unison. True, they disagree as to the timing and ultimate consequences of a warming earth, but they are in virtually unanimous agreement when it comes to defining the trend: warmer.

I suppose, in view of the parade of severe winters in the late-1970s, a statement of that sort seems highly contradictory. It is a point worth exploring.

A COOLING CYCLE

Natural cycles of warming and cooling still dominate our weather and climate. One of the more important cooling cycles in the Northern Hemisphere appears to be one that reaches a nadir roughly every 180 years. That is, about every 180 years a significant drop in atmospheric temperature takes place and persists for several decades. In reality it is not quite that simple; there are cycles superimposed upon cycles superimposed upon cycles, and the cooling around the hemisphere is never uniform (some regions may actually grow warmer). But, by and large, the 180-year cycle seems to be a dominant one.[2]

The bottom of the last cycle was in the early-1800s, which suggests the 1980s will once again bring peak coldness. The rugged winters of 1976–77, 1977–78, and 1978–79, each of which was the coldest on record in some part of the United States, were probably the advance guard of the types of winters we will experience more frequently in the 1980s. In fact, the cold, snowy, and damp winter of 1977–78 may have been the best example of what lies in our immediate future. Not every winter will bring deep snows, floods, and bone-chilling cold, of course, but harsher winters will come somewhat more regularly than they have in the recent past.

This is important to us not only in terms of what will happen to our heating bills—OPEC aside—but also in relation to the greenhouse threat. A natural cooling trend will be likely to mask the initial effects of any human-induced (anthropogenic) warming caused by a build-up of CO_2. Figure 1–1 presents a schematic view of the 180-year cooling cycle (in conjunction with a less important 80-year

Figure 1–1. A schematic depiction of Northern Hemispheric temperatures as suggested by studies of Greenland ice core samples. The implication is that large-scale cooling takes place roughly every 180 years (a less important 80-year cycle is also depicted). If the cycle continues, the 1980s will bring the next marked drop in hemispheric temperatures. (*Weather Watch*, Walker and Company, copyright 1979 by Harold W. Bernard, Jr.)

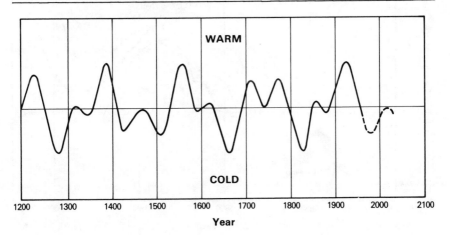

cycle) over the past 800 years. Figure 1−2 gives a more detailed picture, in schematic form, of the cycle since 1800. The cycle is projected out beyond the year 2000.

Figure 1−2 also shows the actual, or measured, temperature of the Northern Hemiphere since the late-1800s. A third curve in the figure

Figure 1−2. A schematic depiction of the 180-year cooling cycle since 1800. Also shown is the actual, or measured, temperature of the Northern Hemisphere since the late 1800s. A third curve estimates what the hemisphere's temperature would be if it were controlled by the CO_2 phenomenon alone. Finally, a fourth curve adds the 180-year cycle and the CO_2 effect together to show what scientists think the hemispheric temperature *should* be. It is apparent that natural temperature cycles are still dominant, but it is also obvious that this may not be the case by early next century! (Reproduced with permission from Broecker, W.S. *Science* Vol. 189, pp. 460−463, Fig. 1, 8 August 1975, copyright 1975 by the American Association for the Advancement of Science. The actual temperature curve has been updated using data from Asakura, and Angell and Korshover. The projection of the actual temperature curve has been added by the author.)

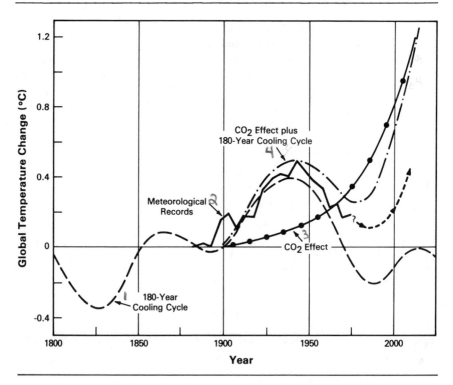

estimates what the hemiphere's temperature would be if it were controlled by the CO_2 phenomenon alone. Finally, a fourth curve adds the 180-year cycle and the CO_2 effect together to show what scientists think the hemispheric temperature *should* be.

As you can see, the measured temperature has remained relatively close to, although somewhat below, the curve researchers think it should be following. What the graph says is that the natural cooling cycle is still having its effect. But it also says that the greenhouse effect possibly is preventing temperatures from dropping as low as they might were they controlled by natural fluctuations alone. Note that before the end of the century, a natural warming trend is projected on the 180-year cycle curve, while at the same time the CO_2 effect really takes off. Thus, shortly after the year 2000 we could be in for a very rude climatic shock, as hemispheric temperatures soar past anything we have experienced in the past several hundred years!

THE IMPACT ON OUR LIVES

By mid-century our climate may be warmer than at any point in the last 125,000 years. The change would come about in a very short period. Change that in the past took hundreds or even thousands of years would happen in just a few decades. It would be an understatement to say humankind would have immense difficulty in adjusting to it.

Following chapters examine in detail, mainly for the United States, the changes in weather patterns likely to result from a warming climate. It is not simply a matter of adding x-number of degrees onto the average temperature for any spot. Global wind circulation patterns also change, and thus weather regimes over particular regions are completely recast. Some spots, in fact, turn cooler. Others become drier, some wetter, and many warmer. Most of the chapters in the book deal with specific climatic changes many of us may live to experience—changes that might occur early in the next century. Chapter 12 presents, in more general terms, a scenario of global climatic transitions that are possible over the next 100 years or so.

Such climatic variations have a direct and significant impact upon our lives. One need only look at recent severe winters to realize how vulnerable economy, commerce, and even politics are to the vagaries of the weather. Agriculture, especially, is highly sensitive to climatic change, and the greenhouse threat holds frightening possibilities for our midwestern breadbasket. They are possibilities that could be realized within just a few decades.

The final chapters suggest ways out of the dilemma, although it is now doubtful that we can completely escape the consequences of what we have started. We can at least attempt to mitigate the threat, while buying time in which to adjust to the changes. In this light the development of solar energy, not the use of coal or the development of synthetic fuels, must receive highest priority. Energy conservation must become a way of life. Above all, the United States needs an energy policy that emphasizes and encourages conservation and the development of solar and solar-related energy sources. A policy that aims at coal and synfuels could be pointing toward climatic disaster.

Michael Glantz, a political scientist at the National Center for Atmospheric Research in Boulder, Colorado, makes the following observation in relation to the CO_2 problem: "In the past the decision-making approach to . . . environmental problems such as air pollution has generally been one of muddling through. That may have been adequate in dealing with some political and social problems, but it has not been an effective way to deal with . . . cumulative environmental problems. Eventually such problems increasingly worsen until a crisis situation is perceived. . . . Then policy making grows out of crisis management."[13]

The greenhouse threat must be met head-on. And it must be met now.

2 WE CAN'T PUT THE WEATHER IN A TEST TUBE

The history of atmospheric carbon dioxide research is short. Continuous measurements of airborne CO_2 were begun in the late 1950s at the observatory on the dormant Hawaiian volcano, Mauna Loa. In 1962, when I was in Alaska measuring soil temperatures in the tundra, John J. Kelley, Jr., a researcher from the University of Washington, was making CO_2 measurements along Alaska's arctic coast.

I thought John was a little weird, putt-putting along the shores of the Arctic Ocean in a clapped-out fishing boat, measuring something as innocuous as carbon dioxide. But thanks to John, and others like him, such as Charles David Keeling—who helped establish the Mauna Loa program—we know for certain that atmospheric CO_2 concentrations have been steadily growing for the past twenty years.

HISTORICAL TRENDS

$\sqrt{}$ CO_2 concentrations have, in fact, probably been growing for over a hundred years, since around 1860 when the industrial revolution began. It was at that time that the use of fossil fuels, coal in particular, started to accelerate. Scientists estimate that the pre-industrial concentration of CO_2 was somewhere between 285 and 305 parts per million (or *ppm*—a chemical measure).

13

√Early measurements by Keeling in the mid-1950s and at Mauna Loa in 1958 pegged the amount at 311 or 312 ppm. Current estimates put the concentration close to 335 ppm, an increase of almost 7 percent since 1958. Researchers think that by the year 2000 the value will be in the range of 380–400 ppm.[1] Figure 2–1 charts the atmospheric carbon dioxide increase as measured at Mauna Loa Observatory between 1958 and 1977.[2] The Mauna Loa record is essentially duplicated by other measurements taken at the South Pole, American Samoa, and Barrow, Alaska, and by readings gathered by Sweden and Australia.

Scientists, by knowing how much fossil fuel we have burned annually, and estimating how much we will burn in the future, can derive a CO_2 curve that covers a much longer time period than that of the Mauna Loa measurements. Figure 2–2 shows their calculations from the beginning of the industrial revolution to the year 2000. The exponential nature of this curve is obvious. The first 10 percent increase in CO_2 took place over about 110 years. The next 10 percent rise will take just twenty years, and the next 10 percent beyond that only ten years.

Figure 2–1. The atmospheric CO_2 content as measured at Mauna Loa Observatory, Hawaii, between 1958 and 1977. The Mauna Loa record is essentially duplicated by other measurements taken at the South Pole, American Samoa, and Barrow, Alaska, and by readings gathered by Sweden and Australia. By 1979 the CO_2 concentration had reached 335 ppm, an increase of almost 7 percent since 1958.

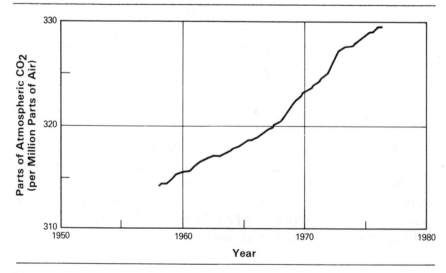

Figure 2–2. Calculated atmospheric CO_2 concentrations from the beginning of the industrial revolution to the year 2000. The exponential nature of the curve is obvious. The first 10 percent increase in CO_2 took place in about 110 years. The next 10 percent rise will take just 20 years, and the next 10 percent beyond that only 10 years. (Schneider, S. "Climate Change and the World Predicament," *Climatic Change* 1 [1977]: 21, Figure 5b.)

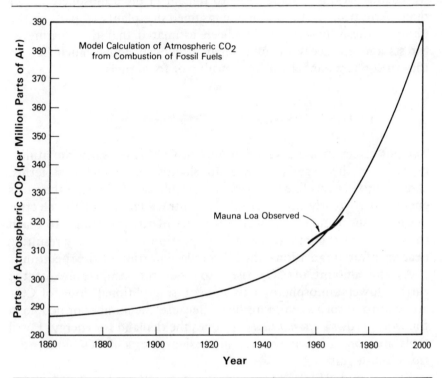

One thing that these CO_2 curves do not show is that there is a seasonal variation in the amount of atmospheric CO_2. That is, superimposed on the steady upward trend of the CO_2 concentrations, there is a seasonal oscillation each year of about 5 or 6 ppm. Because plants and trees use carbon dioxide in the photosynthesis process, when vegetation is actively growing and making carbohydrates, a lot of CO_2 is being used. In winter, when many plants are dormant, less CO_2 is used, and more of it remains in the atmosphere.

So we find that in the Northern Hemisphere atmospheric CO_2 content reaches a peak just after winter, usually in April, and falls to a minimum at the end of summer, usually late September or October.

Plant life then, plays an important part in the CO_2 cycle. As a matter of fact, recent research indicates that the destruction of the world's forests may actually be turning land biota (plants and trees) into a net *source* of CO_2. What that means is that by destroying woodlands we are forcing CO_2 to remain in the atmosphere rather than permitting it to become consumed by plants as part of the photosynthesis process. It has been estimated that the clearing of forests and the decay of humus may effectively add as much CO_2 to the atmosphere annually as the burning of fossil fuels.[3]

FUTURE TRENDS

The net effect of increasing amounts of CO_2 is to warm the atmosphere near the earth's surface. In slightly more technical terms, carbon dioxide acts like the glass in a greenhouse,* allowing the sun's energy, particularly in the visible and ultraviolet end of the electromagnetic spectrum — the short wave part of the spectrum — to reach the earth, but then absorbing a large portion of the earth's resulting heat or infrared radiation — the longer wave portion of the spectrum.

As the amount of CO_2 rises, so does the temperature of the earth's lower atmosphere. This is because additional doses of CO_2 are able to absorb even more heat. Because the number of people in the world grows *exponentially*, so does the demand for energy. Fossil fuels are consumed at an ever increasing rate, and CO_2 concentrations follow suit.

By the mid-1960s, researchers were becoming interested in the effects this rapid CO_2 build-up would have on our climate. In 1967 Syukuro Manabe and Richard Wetherald of the U.S. Department of Commerce's Geophysical Fluid Dynamics Laboratory in Princeton, New Jersey, warned of a global temperature increase on the order of slightly more than 4°F if atmospheric CO_2 were allowed to double.[4] (Current estimates say this will happen sometime between 2020 and 2050.)

It is fair to ask how scientists know how much warming might take place, because unlike the increase in CO_2 itself, CO_2-induced

---

*A greenhouse is effective for other reasons: it physically prevents outside air from entering the greenhouse area.

warming is not something we have been able to measure directly. (This is because of the natural cooling trend that may be camouflaging any CO_2-fostered warming.)

Weather and climate are not entities that can be handily studied in a test tube. Researchers must produce representations of atmospheric behavior using mock-ups, or models, constructed with numbers and complex mathematical relationships. Then, with the aid of computers, the models can be used to examine atmospheric response to various influences, such as increasing amounts of CO_2.

NERVOUS BREAKDOWNS
AND OTHER PROBLEMS

Climate models are relatively primitive. Scientists cannot possibly hope to represent all of the intricacies and nuances of real atmospheric processes. And even today's largest computer would have a nervous breakdown trying to deal with the complexity of such a model. Furthermore, the numerical mock-ups require that various assumptions be made. The assumptions may or may not be right. And the projected numbers that are plugged into the models may or may not be right.

For instance, in the case of the carbon dioxide problem, researchers have a good idea of how much CO_2 is being released into the atmosphere every year. However, since 1958 only about half of the CO_2 known to have been produced by fossil fuel consumption has shown up in the atmosphere. Where has the rest of it gone? Certainly not into the forests and plants if they are indeed acting as a net source of CO_2. Some oceanographers say that though the oceans can absorb a substantial amount of CO_2, they cannot absorb 50 percent of the amount produced by the burning of fossil fuels every year.[5] Thus, modeling is made difficult by such questions.

Even the most complex climate models have serious shortcomings. One of the greatest weaknesses is that generally types and amounts of clouds are not computed. (They are accounted for in an "average" way by adjusting incoming and outgoing radiation.) A second major deficiency is that interactions between the atmosphere and oceans are not represented. Such interactions are important, because the oceans store and transport large quantities of heat and thus exert significant long-term effects on weather and climate. (It should be noted that a major effort to include air-sea interactions in climate modeling has recently been initiated.)

But, as Dr. John Mason, director general of the Meteorological Office, Bracknell, England, points out, ". . . despite these deficiencies, the models successfully stimulate the major features of global atmospheric circulation and of present world climate, at least as far as the averaged conditions are concerned."[6]

Since the late 1960s additional work has been done on the CO_2-climate problem, and more complex models have been employed. But even the more complex mock-ups have given about the same answer that Manabe's and Wetherald's relatively simple model did in 1967. The consensus among greenhouse effect researchers now is that a doubling of the atmospheric CO_2 content will lead to a warming of our climate on the order of 3.6°F to 5.4°F, with significantly larger warming—perhaps as much as 18°F—in the world's polar regions.[7]

The higher warming in the polar regions comes about for a couple of reasons. The first concerns water vapor. Water vapor normally absorbs a great deal of the earth's infrared, or heat, radiation. Thus, in areas where water vapor is fairly abundant, such as in the middle latitudes, CO_2 competes with water vapor for absorption rights. But in polar regions where the air is quite cold and therefore cannot hold very much water vapor, CO_2 has the heat radiation pretty much to itself. So, the addition of CO_2 to areas of the earth that are quite dry—such as the poles (and deserts)—produces a relatively large increase in heat absorption and, therefore, warming. However, in places where there is a lot of water vapor, the absorption and warming effects are noticeably smaller.

The second reason follows the first. Once the warming in the polar regions is under way, and ice and snow begin to gradually disappear (melt), the warming accelerates. This is because the snow and ice previously reflected much of the sun's radiation back to space before it was able to warm the earth's surface; but as the snow and ice melt the water and land become "absorbers" instead of reflectors, and additional warming occurs. The whole process becomes self-enhancing: more warming leads to more ice melting, which leads to more solar radiation being absorbed, which leads to more warming, etc.

FEEDBACK

In considering how well the researchers and models have done their job in figuring out the extent of the warming in store for us, there are a couple of more important problems that must be addressed. These are the problems of *feedback*. For instance, there is the problem of negative feedback. The argument for that goes like this: As increased CO_2 warms the atmosphere, evaporation might be increased (warm air can hold more water vapor) and the result would be more cloudiness. More cloudiness would block more sunshine, and the warming effect of the additional CO_2 concentrations would be negated. Atmospheric cooling might even be triggered. However, at a 1976 international conference on the carbon dioxide problem, held in Berlin, evidence was presented indicating that the feedback between temperature and clouds does not have any significant effect on the earth's warming.[8]

There is also the problem of positive feedback: In this scenario CO_2-induced warming of the lower atmosphere warms the oceans; because warmer oceans can retain less CO_2, they release previously absorbed CO_2 back into the atmosphere where it further accelerates the warming. Eventually, runaway heating of the earth results, happening even sooner than predicted by the models. But recent work on this problem has shown initial estimates of the feedback mechanism were 10 times too large, and that positive feedback does not pose the dangers previously thought.[9]

Still, for all the efforts of the scientists who have studied the CO_2 threat, the answers the models are giving might be wrong. An incorrect assumption might have been made someplace along the line, an atmospheric process incorrectly modeled, or an important process ignored through lack of knowledge. A wrong number, or set of numbers might have been used. Or perhaps the models just are not sophisticated enough to tackle the problem. But—and this is very important—if there is a significant error in the model forecasts, that error could just as well be in the direction of underpredicting the amount of potential warming as in over-predicting it. In fact, Cesare Marchetti, of the International Institute for Applied Systems Analysis, Laxenberg, Austria, thinks that is exactly the case: "The personal opinion of people working in the area . . . is that refinement of the models and consideration of all possible feedbacks will probably force us to draw a bleaker picture."[10]

HOW MUCH PROOF?

A huge amount of work remains to be done on the problem. Of course, by early next century it will become clear whether or not the current future-climate calculations are correct. But by then it will be too late to redirect our energy development programs so that they avoid significant climatic warming.

Dr. Stephen H. Schneider, a research climatologist at the National Center for Atmospheric Research, Boulder, Colorado, articulates the social dilemma we face: "There are always people who will say that we do not know exactly what will happen, and they will always be right, since absolute accuracy can never be obtained in science. . . . The question is: What proofs must be submitted and at what risk can we proceed?"[11]

Schneider has this suggestion: "Since the consequences of a climate change at the higher end of the current [warming] estimate could be both enormous and possibly irreversible, perhaps society would be best to err conservatively in planning future fossil fuel consumption patterns."[12]

Many of the following chapters in this book explore what these "enormous and possibly irreversible" climatic effects might be, beginning with changes many of us may live to experience. We should know what we are facing.

HOW FAST?

The speed with which the climate alterations will occur will depend upon our use of fossil fuels. If we fail to wean ourselves of dependence on the fossils and their synthetic derivatives, and continue to consume them at the historical growth rate of 4 to 5 percent a year, the amount of atmospheric CO_2 will double by the year 2020.[13] If the world's energy growth rate is somewhat less than 4 or 5 percent—which it apparently has been recently and is projected to be—say around 3 percent per year, and if our demand for the fossils levels off early next century, then the doubling of CO_2 will not occur until closer to midcentury.[14] This appears to be the most likely scenario, and the one I will use for the purposes of this book.

Figure 2-3 shows the projected global temperature trend based on that scenario. The figure indicates that by the year 2000 our climate

Figure 2–3. The projected global temperature trend based on the assumptions that (1) the world's energy growth rate averages around 3 percent per year over the next several decades, and (2) our demand for fossil fuel levels off early next century. Under this scenario the world would be 4° or 5°F warmer by the year 2040. (From Jill Williams, International Institute for Applied Systems Analysis; published in: Gribbin, J. "Fossil Fuel: Future Shock?" *New Scientist*, 24 August 1978, p. 541.)

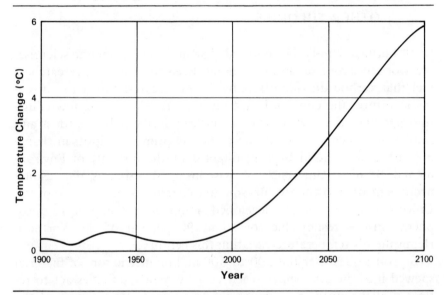

would be about 1°F warmer than it is now. However, the probable masking of the CO_2-caused warming by natural cooling (see Figure 1–2) suggests the transition to the 1°F warmer climate may be slightly delayed, to the period 2010–2020. In the ensuing decade, 2020–2030, the warming should begin to accelerate, reaching 2 to 3°F. Then by 2030–2040, with the CO_2 content of the atmosphere virtually twice what it is now, the warming will be likely to reach the proportions mentioned earlier in this chapter, 3.6 to 5.4°F!

Even assuming the most conservative energy consumption projections—a growth rate of about 3 percent per year or a bit less, and a world in which nonfossil energy sources dominate soon after 2000—the atmospheric CO_2 buildup may well reach 475 ppm by 2050 with a resultant climatic warming of over 2°F.[15]

It is obvious then, given a reasonable amount of certainty in the CO_2-climate models, that if we are to preclude significant humanmade climatic consequences, we must become dependent on non-fossil energy sources within the next decade or two. That, unfortunately, will not happen. But that should not prevent us from applying maximum effort to obtain a fossil-free energy world as quickly as possible. We must if we are to avoid the possible global disaster of runaway warming.

OTHER THEORIES

In all fairness it must be reported that not all atmospheric scientists are worried about runaway warming. Even though most researchers feel that carbon dioxide offers by far the greatest threat to our climatic future, the concern is not unanimous. There are a few climatologists who think that airborne anthropogenic particles (dust and smoke) have been and continue to be the primary culprits in changing our climate. Reid Bryson, director of the Institute of Environmental Studies, University of Wisconsin, Madison, is the leading proponent of that theory. Bryson and Gerald Dittberner, also of the University of Wisconsin, developed a model that indicates atmospheric dust is responsible for about 90 percent of the Northern Hemisphere's temperature variation this century.[16]

Bryson argues that the 500 to 600 million metric tons of material spewed into the atmosphere each year by man's activities act to reflect sunlight away from the earth. The average temperature of the earth, he says, has thus been reduced.

Bryson's assumption, or model, that anthropogenic dust in the atmosphere leads to cooling is not universally accepted. Under certain conditions atmospheric particles may actually act to increase temperatures by absorbing long wave (heat) radiation from the earth—the greenhouse effect again. This situation would most likely arise over land areas, especially those with a high reflectivity, such as the ice- and snow-covered polar regions.[17]

Dr. Hurd Willett, professor emeritus of Meteorology at the Massachusetts Institute of Technology (MIT), Cambridge, points out that the most extreme cooling in recent decades has been in the latitudes of 60°N to 80°N. That is north, he says, of where the most pronounced cooling induced by humanmade dust would be expected.[18] (If anthropogenic dust were the main cause of the recent cooling trend, maximum effects should have shown up in the industrialized midlatitudes, say between 35°N and 55°N.)

Another human activity that has caused concern in relation to changing our climate is the direct input of heat to the atmosphere. All energy ultimately degrades into heat. We produce heat when we heat or cool our homes and businesses, when we drive our cars, when we manufacture steel, when we turn on the lights at Yankee Stadium. Any process that uses energy releases heat into the environment.

Not only do our activities directly inject heat into the atmosphere but heat is also injected as an indirect result. Energy conversion efficiencies are typically quite low—about 30 percent for electric power plants—so up to 70 percent of the heat ultimately generated is released as waste heat right at the plant site. Even at that, the total amount of heat released by all human activities is currently only about 0.01 percent of the total solar radiation that reaches us. The effects, therefore, of anthropogenic heat are well below the level that can be detected or calculated.[19]

Perhaps, in a few thousand years, we will have created so much thermal pollution that the earth's surface will be about as hot as the sun's. Well, that obviously will not happen, because long before that we "would be wisps of glowing plasma," in the words of Frederik Pohl.[20] The point is, if we weather the greenhouse threat, we will eventually have to tackle the problem of waste heat.

Or maybe not, maybe the long awaited and much publicized "return of the Ice Age" will save us from our own hand.

THE ICE AGE COMETH?

Certainly the threat of another ice age was the topic of much scientific and popular discussion in the 1970s. Books and articles entitled "The Cooling," "Blizzard," "Ice," and "A Mini Ice Age Could Begin in a Decade," abounded. The "snow blitz" theory was popularized on the public television presentation of "The Weather Machine" in 1975. And certainly the winters of the late 1970s were enough to send shivers through our imaginations.

It's true! We probably are heading back toward an ice age. It will not be much of a problem for about another 10,000 or 20,000 years, however. If our planet is still around then, we should have plenty of time to prepare.

In the 1930s a Serbian geophysicist, Milutin Milankovitch, developed a theory that related global climate changes to the earth's orbital behavior. Basically, what Milankovitch did was postulate that the advance and retreat of ice ages were dependent upon variations in the seasonal and latitudinal distribution of solar radiation. He argued that these variations were caused by changes in the earth's orbital geometry.

Other researchers through the years either criticized or attempted to improve upon his work. But they, like Milankovitch, were hampered in their efforts by lack of an adequate global climate record covering a period long enough to prove or disprove their ideas. However, recent research by a team of American and British scientists finally verified the Serbian's theory.

John D. Hays of the Lamont-Doherty Geological Observatory, Columbia University, is head of the CLIMAP (Climate: Long-range Investigation, Mapping, and Prediction) project on which the team worked. Hays says, "We are certain now that changes in the earth's orbital geometry caused the ice ages. The evidence is so strong that other explanations must now be discarded or modified."[21]

By studying sediment cores from a relatively undisturbed section of the Indian Ocean floor, the CLIMAP project was able to uncover an unbroken geological record of climate covering more than 450,000 years. Modern dating techniques—including the use of radioactivity measurements, knowledge of the evolutionary history of microscopic, fossilized animals, and samples of volcanic ash and magnetic reversals of known ages—made possible a chronology three times longer than any previous one, at least one dated so accurately.[22]

Close examination of the cores—for the fossilized remains of plants and animals that thrived during different climatic regimes—revealed significant climatic cycles of 23,000 years, 42,000 years, and 100,000 years. These cycles are nearly the same as the overlapping cycles of variations in the precession and tilt of the earth's axis, and in its orbital eccentricity.[23]

The precession, or "wobble," of the earth's axis in its eliptical orbit around the sun means that over a period of centuries the earth is closer to the sun at different times of the year. Currently the earth is closest to the sun in January. In about 10,000 years it will be closest in July. A complete precession has a cycle of about 21,000 years. Other things being equal, a greater summertime distance between the earth and sun should mean cooler temperatures and less melting of the polar ice caps (i.e., ultimate ice cap growth over a period of years).

The earth's orbit also tilts (in relation to the sun), and the amount of tilt varies over the centuries; thus the earth's axis is sometimes closer to perpendicular to the sun's radiation than it is at other times. The maximum deviation from perpendicular occurred around 9000 years ago. Now the trend is toward a period of minimum tilt. A complete cycle of maximum to minimum and back to maximum tilt takes 41,000 years. Times of minimum tilt are conducive to ice ages. This is because at minimum tilt the earth's poles are more nearly equidistant from the sun all year, and there is a minimum of seasonal (i.e., winter to summer) effects. Much of the polar ice never melts and the ice caps slowly build. During periods of maximum tilt, one pole, the one in the summer hemisphere, receives continuous solar radiation, and the ice caps tend to retreat.

The eccentricity of the earth's orbit also changes with time. That is, over a period of about 100,000 years it changes from nearly circular to more eliptical. The more eliptical path may act to accentuate the impact of the variations produced by the seasonal differences in distance between the earth and sun. For instance, when the orbit is more nearly circular, the precession effect (the 21,000-year cycle) has less influence because the seasonal differences in distance between the earth and sun are relatively small. When the eccentricity is at a maximum, the seasonal differences produced by the precession are also at a peak. Thus, the eccentricity cycle may be the dominant one of the three. The earth's orbit is now about halfway between the two extremes of the 100,000-year cycle and is moving toward greater eccentricity.

Hays and his coworkers extrapolated the orbital cycles and put them into a climate model. "The results indicate that the long-term trend over the next 20,000 years is toward extensive Northern Hemisphere glaciation and cooler climate." They issue the caveat that the prediction ignores anthropogenic effects.[24]

In view of the greenhouse threat, that is a pretty significant caveat.

3 A SEARCH FOR A CLIMATIC ANALOG

Numerical models of the atmosphere can give scientists a broad, generalized idea of how our climate might change under a given set of circumstances. But the models cannot be very specific. For instance, the models can tell researchers that additional amounts of atmospheric CO_2 lead to a warming earth, but they cannot define which geographical regions will warm (or cool) and to what extent, and they cannot indicate in what manner precipitation patterns might be altered. This is because the models are not yet adept at predicting atmospheric circulation patterns over a long period—months or years. As the temperature of the earth's atmosphere varies, so do the accompanying circulation patterns. It is these circulation patterns that must be closely examined if we are to learn the details of how our climate may change under the influence of the greenhouse effect.

THE WESTERLIES

Climate regimes are determined by the configuration of the earth's wind patterns. The wind, or circulation, pattern that is most important, at least for the purposes of this discussion, is the one commonly known as the *jet stream*. The jet stream is actually only a small part of the westerlies, the name given an endless band of upper atmospheric winds that circle the hemisphere at midlatitudes (30°−65°N).

The jet stream is the axis of the strongest westerlies. It usually howls along at around 30,000 feet and may reach speeds up to several hundred miles per hour. In general, the westerlies are the reason a Boeing 747 can fly from Los Angeles to New York faster than it can from New York to Los Angeles.

In reality, the westerlies do not blow from west to east all of the time. More typically they display an undulating pattern, sweeping southward over one part of the hemisphere, and back northward somewhere else. Occasionally they may form a complete loop. The undulations and loops constantly change positions, and it is these shifts in the configuration of the westerlies that produce changes in our weather regimes.

Where the westerlies shift or loop northward they bring relatively warm, dry weather. Where they bend southward, generally cooler, wetter conditions result. Figure 3−1 presents a schematic representation of the westerlies and how they affect our weather.

As the earth's climate warms or cools over a long term there is a concomitant shift in the preferred pattern of the westerlies. Not only do the positions of the loops and swirls change, but the actual number of such undulations may also change. Figure 3−2 gives an example. During a historical period known as the Little Ice Age, which culminated in the 1600s, the atmospheric circulation pattern was markedly different from what it was during an earlier (warm) era in the Middle Ages from about 1000 to 1200.

The jet stream during the Medieval Warm Period was farther north than during the Little Ice Age, and−in the summer−displayed four basic southward bends as opposed to five during the Little Ice Age.[1] The southward dip over the British Isles and western Europe during the Little Ice Age (see Figure 3−2) implies that much cooler weather prevailed there at that time. Indeed, that was a time during which Alpine glaciers expanded to their maximum extent in historical times, burying pasturelands and passes in great sheets of ice. The shores of Iceland, now ice-free except for a few months each year, were locked in ice for up to six months every year.[2]

During the warm period that preceded the Little Ice Age the weather in Europe was extraordinarily good. Vegetation and glacier boundaries were roughly 500 to 600 feet higher than even today. Grapes were grown in England and East Prussia, where late frosts

Figure 3−1.   The undulating pattern of the westerlies. A looping (or blocking) pattern is indicated over western Europe. Such a configuration would produce fair weather over Scandanavia, and cloudy, showery weather in Spain and France. The pattern shown over North America would produce warm, dry weather in the western United States, and cool, wet conditions in the East. (*Weather Watch*, Walker and Company, copyright 1979 by Harold W. Bernard, Jr.)

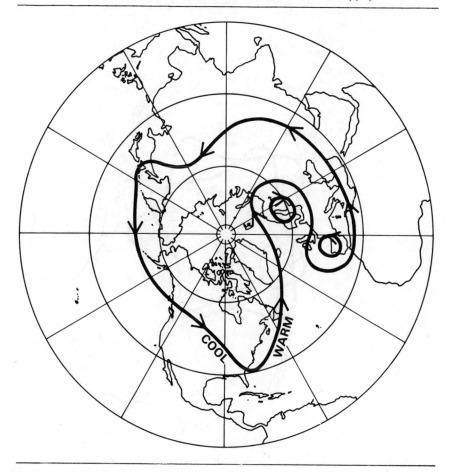

Figure 3-2. The jet stream—the axis of the strongest westerlies-during the Medieval Warm Period (pattern A) was farther north than during the Little Ice Age (pattern B), and—in summer—displayed four basic southward bends as opposed to five during the Little Ice Age. The southward dip over the British Isles and western Europe during the Little Ice Age implies that much cooler weather prevailed there then.

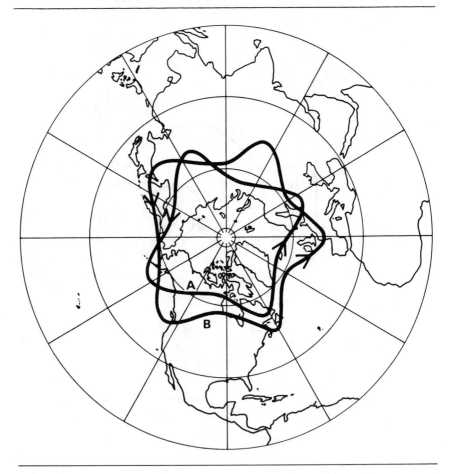

now preclude vineyards; in southwest Greenland sheep and wood-lands, not ice, prevailed.[3]

Our contemporary climate is somewhere between that of the Little Ice Age and the mild Middle Ages. Our summer circulation pattern is intermediate to the four and five loop systems depicted in Figure 3−2.[4] There are some indications that over the next couple of decades the configuration of the westerlies will be closer to that of the Little Ice Age than to that of the Middle Ages,[5] but it is really the period after the next twenty to thirty years that we want to explore. We want to discover what happens when the greenhouse warming begins to take hold.

We already know this much: even during times of significant climatic warming, such as took place throughout the Middle Ages, not all regions in fact grow warmer. For instance, during the Medieval Warm Period, much of western Russia was probably relatively cool.[6] This is suggested by the southward dip in the jet stream there, shown on the schematic diagram in Figure 3−2. This is why it is important to determine what type of atmospheric circulation pattern will accompany an expected climatic change; regional differences can be clearly defined. Because numerical models of the atmosphere cannot provide such specificity, we must turn to past climatic regimes to discover what lies ahead. As Dr. William W. Kellogg, National Center for Atmospheric Research, Boulder, Colorado, points out, "Using the real Earth for a model is at least as good as, and probably better than, the theoretical numerical models."[7]

Another view on the matter comes from Australian climatologist Dr. Garth Paltridge: "I have the feeling that this sort of prediction will come not directly from mechanistic models, but rather from a careful analysis (and interpretation in the light of model behavior) of the correlations and persistences in past records of the observed weather and climate."[8]

What Dr. Paltridge is saying is that we have a better chance of predicting our climatic future by correlating expected trends with past behavior. In the case of a warming climate we want to find a parallel, or *analog*, in historical weather records and attempt to draw some conclusions about the future using the analog as our model. The best way to do this is by examining data from a period in which regular, scientific, and comprehensive weather observations were made.

Records from the Middle Ages and the Little Ice Age are few and far between. Professor Hubert H. Lamb of the Climatic Research

Unit in England was able to determine the probable configuration of the upper winds associated with those periods by studying the sparse records of surface temperature and atmospheric pressure that are available. Then by applying a deductive process involving knowledge of the various vertical relationships among temperature, pressure, and wind, he was able to reconstruct what the high-level wind patterns of those times most likely were.[9] Fortunately we do not have to journey that far back in time to find a good analog for the initial stages of a warming earth.

DUST BOWLS AND ICE PACKS

The 1930s marked the peak of the most recent global warming. Figure 3–3 shows Northern Hemisphere temperature trends since 1880. The warming trend into the late 1930s is readily apparent, as is the cooling of the hemisphere since that time. Dr. Hurd Willett points out that concomitant with the warming trend the westerlies reached their maximum northward extent in the 1930s. After that they began to display a greater frequency of looping (or blocking) patterns, and we went into what Willett terms a period of *climatic stress*, which lasted through the 1950s.[10] I will talk about climatic stress in a later chapter.

Figure 3–3.   Northern Hemisphere temperature trends since 1880. The warming trend into the late 1930s is readily apparent, as is the cooling of the hemisphere since that time. (Updated from *Weather Watch*, Walker and Company, copyright 1979 by Harold W. Bernard, Jr.)

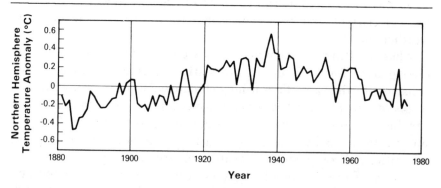

The 1930s were years of sweltering heat waves, earth-cracking droughts, and maximum tropical storm frequency. In the United States a greater number of state records for high temperature and dryness were set during the 1930s than during any other decade since the 1870s.[11] A superdrought turned the Midwest into a vast dust bowl, and people by the thousands fled the area. Russian droughts triggered the liquidation of the Kulaks, and severe drought stalked southern Australia.[12]

In the Atlantic, the Caribbean, and the Gulf of Mexico, tropical cyclone frequency reached a modern high. Twenty-one tropical storms and hurricanes blew up in 1933, and seventeen in 1936. The current (1946–1975) average is nine.

As the climate warmed, so did the earth's oceans. In the seas around Greenland species of cod common to the southern oceans migrated northward almost 10° of latitude between 1913 and 1930. In 1930 they appeared in large numbers off Greenland's west coast near Disko Bay at about 70°N, causing a considerable rise in the fishing activity there. Cod also appeared in commercial quantities off the coasts of Norway, Spitzbergen, Iceland and in the Barents Sea north of Russia.[13]

In 1935 a Soviet icebreaker sailed north in open water from an island (Novaya Zemlya) in the Arctic Ocean to 82°41'N. In 1901 a much more powerful icebreaker had been unable to reach the northern tip of Novaya Zemlya near 77°N. Today, year-round pack ice is usually found from 80°N northward. In 1938 another icebreaker cruised north from Siberia and reached 83°05'N, the most northerly point attained by a ship sailing freely. Today in that region the Arctic Ocean is generally covered by unnavigable ice north of 77°N. During the 1930s there was also a rapid retreat of glaciers in Iceland, Spitzbergen, Franz Josef Land, and other islands of the Arctic Basin.[14]

The 1930s were indeed warm, and will provide our climatic analog for the initial decade of the CO_2-induced warming that I expect may occur between about 2010 and 2020. By that time our climate could be about 1°F warmer than it is now. The 1930s averaged about 1°F warmer than recent decades have. However, to reiterate, it is not a matter of just adding 1°F onto current normal temperatures and saying that is what to expect. Records from specific areas must be analyzed in detail to determine how the varying atmospheric circulation patterns affected different regions. The answers we obtain from examination of the 1930s weather records will give us a good indication of what to expect early next century if anthropogenic warming becomes a reality.

We will start with what is perhaps the most frightening aspect of the greenhouse threat: drought.

4 THE RETURN OF THE DUST BOWL

Drought is an extended period of time with inadequate precipitation. In an agricultural sense, however, drought does not necessarily begin with the cessation of rainfall. It begins when the available stored water in the soil cannot meet the evaporative demands of the atmosphere.[1] Heat and wind contribute to evaporation and hence to drought.

DROUGHTS AND SUNSPOTS

On a broad time scale the occurrence of major droughts in the United States west of the Mississippi River appear to follow a fairly regular cycle of twenty to twenty-two years, and to match up very closely with the double sunspot cycle, two eleven-year cycles. Sunspots are relatively small, dark areas that appear on the sun's surface. In reality they are great magnetic storms that swirl through the hot, gaseous *photosphere* of the sun. Sunspot activity progresses through an irregular cycle of roughly eleven years during which the number of spots moves from a minimum to a maximum and back to a minimum. The number of actual sunspots may be as low as zero or as high as two hundred, on an average annual basis. And the cycle may vary in length from ten to fifteen years. As Figure 4–1 shows, major

Figure 4-1.   Annual smoothed sunspot numbers with the peak years of major western droughts indicated by dots. These major droughts have occurred in conjunction with every other sunspot minimum since 1700. Extrapolation would suggest the next important drought in the West will occur around the year 2000, with the next one after that due about 2020—in conjunction with the greenhouse effect! (*Weather Watch*, Walker and Company, copyright 1979 by Harold W. Bernard, Jr.)

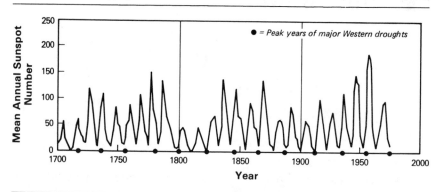

droughts in the western United States have occurred in conjunction with every other sunspot minimum back to 1700.

In this century the drought that led to the dust bowl days of the 1930s is part and parcel of American history; severe drought stalked the Great Plains again in the mid-1950s, and California and the Rocky Mountains in the mid-1970s.

By extrapolating the double sunspot cycle, meteorologists can say with some certainty—although not absolute certainty, for cause and effect is still a mystery—that the next major drought in the West will occur around the year 2000. What cannot be forecast is *exactly* when the drought will occur or precisely where it will occur. Some researchers feel that the drought due around the turn of the century will be centered in the southwestern United States, with southern California, Arizona, and New Mexico suffering the brunt.[2] But it is not that drought we are concerned with in this book. It is the one likely to strike after that, probably around the year 2020, near the end of the first decade of significant CO_2-caused climatic warming. That drought may well have all of the earth-cracking dust bowl characteristics of our climatic analog period, the 1930s.

BLACK BLIZZARDS

√The greatest disaster in American history attributable to meteorological factors was the superdrought of the 1930s, which dried up 50 million acres of the Great Plains.[3] Poor agricultural practices and overuse of the prairies contributed to the catastrophe. Before the plains were settled in the late 1800s, natural, deep-rooted grasses could survive even prolonged droughts. But after 1885 the grasses were either plowed under to raise wheat and corn, or were overgrazed by range cattle. (Drought, blizzards, and overproduction collapsed the Cattle Kingdom in 1887.) The midwestern topsoil was bared to the mercy of the elements, and the relentless winds that accompanied the 1930s drought removed 350 million tons of the richest soil in the world.[4]

The first fingers of the dust bowl drought crept into the Great Plains in late 1930, then stretched out through early 1931 to grasp the northern and central plains. Thereafter, every year through 1939 brought serious drought to some part of the Midwest. In 1934 and 1936 the entire region from Texas to Canada was scourged. Dendrohydrologists (scientists who study climate records as revealed in tree rings) have determined that the driest single year in the western United States since 1700 was 1934.[5]

The most seriously affected areas were in the southcentral plains, however. The contiguous parts of Colorado, New Mexico, Kansas, Oklahoma, and Texas were the real dust bowl. From 1934 through 1936, Dalhart, Texas, near the center of the region, averaged just 11.08 inches of precipitation per year, or 58 percent of normal. Between August 1932 and October 1940 the moisture index in western Kansas was continuously below normal, with 38 months registering extreme drought.

The first of a series of awesome dust storms, or "black blizzards," struck in November 1933. Vast quantities of dust were swept thousands of feet into the air from Montana to the western Ohio Valley. Dust reached the Atlantic seaboard from Georgia northward; "black rain" fell in New York state, and "brown snow" in Vermont.

Great dust storms continued to ravage the Midwest until 1939. In 1934 and 1935 skies were darkened over the Ohio Valley and Great Lakes. Sunshine was dimmed in the eastern United States and dust particles sifted down on ships in the Atlantic Ocean.

In the heart of the dust bowl, livestock died of suffocation and starvation; what crops survived were withered and stunted. Huge drifts of loose soil surrounded farms and blocked highways and railroads. The blowing dust sandblasted paint off houses and automobiles. Hundreds of people died of respiratory ailments, and thousands fled the area, simply abandoning their ranches to the wind and dust. The black blizzards left profound physiological and psychological impacts on a generation of midwesterners.

SNAPSHOTS

The extent and severity of the dust bowl drought can be compared to the two most recent major western droughts by examining Figures 4–2, 4–3, and 4–4. They are based on the Palmer Index, a measure

Figure 4–2. A "snapshot" of the 1930s drought at its peak in July 1934. At this point the most extreme dryness had shifted somewhat north of the heart of the dust bowl, and stretched from coastal California to the eastern Great Lakes. Centers of very intense drought could be found in Nevada, Idaho, Utah, Minnesota, and Illinois.

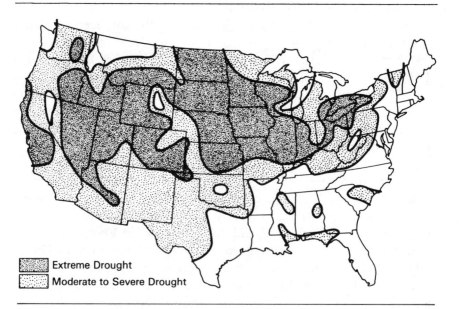

☐ Extreme Drought
☐ Moderate to Severe Drought

Figure 4–3. A "snapshot" of the 1950s drought near its peak in July 1956. This drought was shorter, more confined, and centered a bit farther south than its 1930s predecessor; areas from Nebraska through Texas suffered the most.

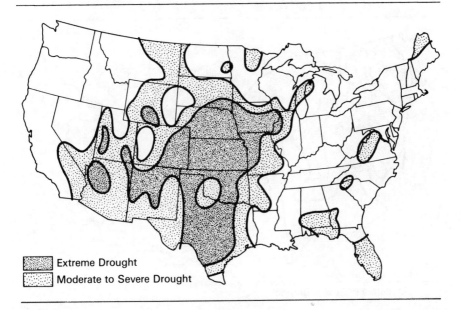

Extreme Drought
Moderate to Severe Drought

of meteorological drought calculated from temperature and precipitation data as recorded by instruments.

Figure 4–2 is a "snapshot" of the 1930s drought at its peak in July 1934. At this point the most extreme dryness had moved somewhat north of the heart of the dust bowl, and stretched from coastal California to the eastern Great Lakes. Centers of very intense drought could be found in Nevada, Idaho, Utah, Minnesota, and Illinois.

The drought of the 1950s was shorter, more confined, and centered a bit farther south than its 1930s predecessor; areas from Nebraska through Texas suffered the most. Figure 4–3 is a picture of drought conditions in July 1956.

The 1970s drought spared much of the Midwest, but dried up reservoirs and snowpacks in the Far West and Rocky Mountain states, and hit hard at Minnesota and Wisconsin. Figure 4–4 is a snapshot of the drought in April 1977, as the western United States came out of

Figure 4–4. A "snapshot" of the 1970s drought in April 1977, as the western United States exited from a virtually snowless winter. This drought spared much of the Midwest, but dried up reservoirs and snowpacks in the Far West and Rocky Mountain states, and hit hard at Minnesota and Wisconsin.

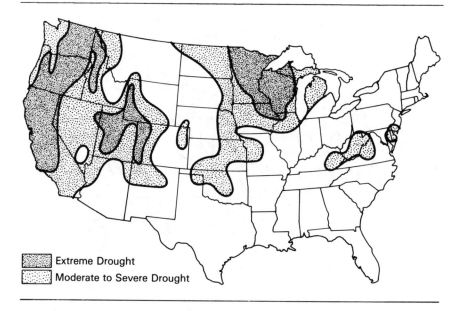

Extreme Drought
Moderate to Severe Drought

a virtually snowless winter. Very intense drought also gripped northwest Wisconsin.

For all their value, though, scientific descriptions of a phenomenon seldom carry the mood of an event. Palmer Indexes and precipitation figures can be analyzed, computerized, and scrutinized, but the true story, the true feeling of a drought period will never emerge. Perhaps Nobel Prize winning author John Steinbeck, in *The Grapes of Wrath*, conveys the sense of the great dust bowl better than statistics or graphics ever could.*

The wind grew stronger, whisked under stones, carried up straws and old leaves, and even little clods, marking its course as it sailed across the fields. The air and the sky darkened and through them the sun shone redly, and there was a raw sting in the air. During a night the wind raced faster over the land, dug cunningly among the rootlets of corn, and the corn fought the wind

*Reprinted with permission of Viking Penguin, Inc. from *The Grapes of Wrath*, by John Steinbeck. Copyright 1967 by John Steinbeck.

with its weakened leaves until the stalk settled wearily sideways toward the earth and pointed in the direction of the wind.

The dawn came, but no day. In the gray sky a red sun appeared, a dim red circle that gave a little light, like dusk; and as that day advanced, the dusk slipped back toward darkness, and the wind cried and whimpered over the fallen corn.

Men and women huddled in their houses, and they tied handkerchiefs over their noses when they went out, and wore goggles to protect their eyes.

When the night came again it was black night, for the stars could not pierce the dust to get down, and the window lights could not even spread beyond their own yards. Now the dust was evenly mixed with the air, an emulsion of dust and air. Houses were shut tight, and cloth wedged around doors and windows, but the dust came in so thinly that it could not be seen in the air, and it settled like pollen on the chairs and tables, on the dishes. The people brushed it from their shoulders. Little lines of dust lay at the door sills.[6]

Steinbeck's vivid scenes may exist only in books both now and in the future, for modern soil conservation practices should preclude the return of the great, dark storms of the 1930s. But the specter of severe drought still hovers over the western United States, and the greenhouse threat may well be converted into dust bowl reality if the warmth and atmospheric circulation patterns of the 1930s return early next century. The implications for agriculture and water supply are serious.

EIGHTY-FIVE BUSHELS PER ACRE

Corn accounts for almost 53 percent of the total U.S. grain production. It is a crop highly dependent upon July precipitation and temperature. Typically, the highest per acre yields come in cooler-than-normal summers. In such summers the corn stores more photosynthate (food), and there is usually more rainfall.

During the period 1930–1939, July temperatures in the corn belt states of Iowa, Missouri, Illinois, Indiana, and Ohio averaged from 2 to 4°F or more above modern means. July precipitation averaged just 50 to 85 percent of current normals. The hottest and driest weather blistered Iowa and Missouri, and corn crops withered and died.

Since the 1930s agricultural technology has made immense strides, and crop yields have steadily increased. But even modern technology

Figure 4–5. A plot of simulated corn yields using 1973 technology and harvested acreage. Contemporary yields in this simulation have averaged around 105 bushels per acre (for Iowa, Missouri, Illinois, Indiana, and Ohio). But during the worst years of the dust bowl, yields would have dropped to near 85 bushels per acre, a reduction of about 20 percent! (Reproduced from *Climate and Food*, 1976, with the permission of the National Academy of Sciences, Washington, D.C.)

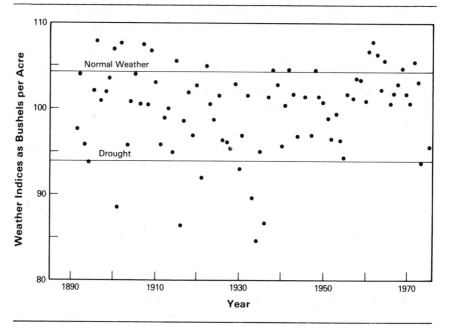

may not be able to overcome the deleterious effects of severe drought. Figure 4–5 is a plot of simulated corn yields using 1973 technology and harvested acreage. Contemporary yields in this simulation have averaged around 105 bushels per acre (for Iowa, Missouri, Illinois, Indiana, and Ohio). But during the worst years of the 1930s dust bowl, yields would have dropped to around 85 bushels per acre, a reduction of about 20 percent![7] Thus, despite modern fertilizers and pesticides, farm mechanization, and strains of grain that are relatively drought and pest resistant, an intense drought would still have a huge negative impact on midwestern corn production.

WHEAT, RUST, AND PRICKLY PEARS

Wheat is the second most important U.S. grain and comprises about 17 percent of our total grain production. There are seven different monthly mean precipitation and temperature variables that significantly influence wheat yields. July rainfall and temperature are again part of the yield equation. During the dust bowl days the average July temperature in western Kansas, the nation's largest wheat-producing region, rose from just under 79°F to over 82°F. And mean July precipitation diminished by 57 percent (from about 2.76 inches to 1.57 inches). The dryness stunted wheat growth, and the hot weather favored wheat rust, a fungus that throughout history has probably caused greater damage than any other disease to man's chief food crop. Wheat rust erupts from rusty pustules on wheat leaves and pushes the epidermis aside. Or, as the wheat matures, black rust blossoms on the stems and sheaths. The red spores of the pest, borne by hot winds, soared over the wheat belt during the 1930s and millions of tons of wheat fell victim. The greatest losses were in 1935—3.4 million metric tons—and in 1937—2.1 million metric tons.[8]

Other pests took advantage of the climate change. The pale western cutworm thrives in warmer and drier conditions. During the 1930s the cutworm population exploded and destroyed thousands of acres of wheat in the Canadian prairies and Montana. In 1937 losses in Alberta ranged from 10 percent to total destruction of individual crops. When the rains returned in the 1940s, the cutworm numbers dwindled.[9]

While the cutworm proliferated in the north, the prickly pear cactus moved eastward out of Colorado during the great drought. The thorny pest invaded 1.6 million metric tonsworth of western Kansas pastureland, and cattle production suffered greatly. But the rains of the 1940s forced the cactus weed back to its Colorado home, and Kansas ranges once again produced sleek cattle.[10]

Our ability to deal with agricultural pests has improved tremendously since the 1930s. But as a 1976 National Academy of Sciences report warns: "Our ability to solve a new problem—related strictly to climatic change and perhaps coupled with emergence of a new pest—in time to forestall disaster is questionable."[11]

Some computer simulations have been done that suggest wheat yields in Kansas, under the influence of a 1930s-type climate, would diminish an average of 3 or 4 bushels per acre over a ten-year period — any one individual year could be significantly worse of course. The 3 or 4 bushels represent about a 15 percent reduction in yield. However, Dr. Louis M. Thompson, dean of agriculture at Iowa State University, points out that while the results of the computer runs indicate the direction of the trends induced by weather effects, "... the coefficients underestimate the effects ..." [12] In other words, in reality the crop yield reductions would be greater.

It is interesting to note that the same computer simulations suggest that despite an increase in temperature and a reduction in precipitation, wheat yields in Illinois and Indiana under 1930s conditions might not change very much or might actually increase. This is mainly because those regions currently have more than enough rainfall to support wheat.

THE OGALLALA AQUIFER

Of course, in the big wheat producing states such as Kansas and Nebraska, we can look for irrigation to ameliorate the effects of future droughts ... can't we? Maybe not.

Much of the irrigation in the wheat belt comes from water in the Ogallala Aquifer, a vast underground reservoir extending from southern Nebraska to the Texas panhandle. This subsurface lake, which is not recharged by surface water, provides for 6.5 million irrigated acres in Nebraska, and 6 million in Texas.

But heavy use of the Ogallala for over thirty years has taken its toll. In some places the water table of this great natural reservoir has dropped 700 feet and is continuing to fall at a rate of 2 to 7 feet per year. Some of the 70,000 wells around Lubbock, Texas, have already dried up. Farms have been abandoned or converted back to dryland operations with an attendant 20 percent drop in production. In Nebraska, dry weather in the mid-1970s resulted in more than 5,000 new wells being drilled in a two-year period.

The depletion of the Ogallala continues, and James E. Osborn, professor of agriculture economics at Texas Technological University, Lubbock, foresees a significant decline in irrigated farming. "It'll happen over the next twenty-five to thirty years," he predicts. "There'll be a slow 35 to 40 percent decline in dollars of production. It is going to happen. . . . It's just a matter of time." [13]

THE RESILIENT CROP

Soybeans, which account for about 14 percent of U.S. grain production, which in turn accounts for 60 to 70 percent of all soybeans grown in the world, are—like corn and wheat—affected by July precipitation and temperature. But unlike the other grain crops, they can recover from July dryness given adequate rainfall in August.[14] Illinois is the soybean production leader, and August precipitation in and around that state was somewhat greater than normal through the 1930s. Thus, soybeans might turn out to be the most resilient grain crop when dust bowl-type weather returns.

Still, the outlook for Midwestern agriculture under such weather conditions is anything but bright, as we have seen. Paul and Anne Ehrlich, in *The End of Affluence*, take the forecast out of the realm of statistics: "A lasting midwestern drought, difficulty in importing what we need, and increased demand created by foreign purchases could combine to make food shortages a reality in the U.S."[15] Chapter 12 will discuss the serious problems of feeding the *world's* ever-growing population.

THE TILLAMOOK BURN

An agricultural crop of another kind suffered unimaginable destruction in the early 1930s. Virtually no rain fell on northwestern Oregon in July 1933, and by early August the forests were tinder boxes. On August 14th, sparks from logging equipment touched off a fire in the Gales Creek Canyon area of the Tillamook Forest. Thousands of men fought the blaze for ten days and managed to "confine" it to 40,000 acres. On the eleventh day a sudden shift of wind caught the weary firefighters off guard, the blaze "blew up," and 200,000 additional acres of virgin timber were burned. The conflagration finally burned out, but not before devouring an estimated 2 billion board feet of lumber, or nearly as much as all the lumber turned out by U.S. sawmills the previous year. The atmospheric pollution was immense; ash accumulated to depths of a foot or more in some spots. Economic losses were tagged at between 200 and 350 million depression dollars.[16]

Twenty years after the blaze, the destruction was still apparent. As a child I remember driving through the Tillamook Burn with my parents. A desert of blackened spires stretched as far as the eye could see, ghostly guardians of the young, dwarf forest beneath them.

Drought contributed to a second disastrous Oregon forest fire in 1936. Thirteen people were killed and hundreds of buildings were destroyed in small communities. In 1937, flames raced through timberland around Cody, Wyoming, killing fourteen people and injuring fifty.

WILL WE HAVE ENOUGH WATER?

But agricultural problems are not the only ones that we would face during a prolonged (or permanent) drought in the western United States early next century. Such a phenomenon would also exacerbate what will likely be critical water supply shortages. By the year 2000 only three of the eighteen federally designated water regions on the U.S. mainland are expected to have adequate water supplies: New England, the Ohio Basin, and the South Atlantic-Eastern Gulf area.[17] And these areas may have problems with water purity. The Water Resources Council, an independent federal agency, estimates that even in the absence of drought serious water shortages may plague Arizona, California, Florida, Kansas, Nebraska, Nevada, New Mexico, Utah, Wyoming, and parts of the Pacific Northwest.[18]

Let us consider the water-related turmoil the return of dust bowl-type weather might bring to just one area of the country, that region of the West supplied by the Colorado River Basin. The Colorado River Basin already has difficulties. With the exception of the deserts of the Great Basin, the Colorado Basin has the greatest water deficiency (average precipitation less potential evapotranspiration—soil water loss from evaporation and from transpiration of water by plants) of any basin in the forty-eight adjacent states. Yet more water is exported from the Colorado Basin than from any other river basin in the United States.[19]

The water, tapped from rain and snow spilling off the western slope of the Rocky Mountains, is dammed and drained at numerous spots along its thousand mile Colorado River journey from Colorado to the Gulf of California. It is used to irrigate crops, keep lawns and golf courses green, and fill swimming pools in Beverly Hills. Southern

California relies almost exclusively on water imported from the Colorado River and from precipitation runoff in the Sierra Nevada mountains of northern California. Metropolitan Los Angeles exists thanks to water from the Colorado River, much of which is stored in Lake Powell behind the Glen Canyon Dam in northern Arizona.

The Glen Canyon Dam marks the boundary between the Upper and Lower Colorado River Basins. The Upper Basin, where most of the precipitation falls, largely comprises eastern Utah, western Colorado, southwest Wyoming, and a small part of northwest New Mexico. The Lower Basin is predominantly Arizona and the adjacent parts of New Mexico, Nevada, and California.

By law the Upper Basin must deliver 75 million acre-feet* (maf) of water to the Lower Basin every ten years, or an average of 7.5 maf per year. An additional 7.5 maf may be required every ten years to help satisfy Mexican water claims. The point where the delivery flow to the Lower Basin is measured is at Lee Ferry, Arizona, just below the Glen Canyon Dam, and the long-term yearly average flow at Lee Ferry is estimated at 14.8 maf. During the great drought of the 1930s there was a significant reduction in precipitation over the Upper Basin (Salt Lake City just to the west of the Basin recorded only 81 percent of current normal precipitation between 1930 and 1939), and the average annual flow at Lee Ferry dwindled to 11.8 maf, or 118 maf over a ten-year period. Still 118 maf would appear to be plenty of water to satisfy the potential 82.5 maf ten-year demand in the Lower Basin. (The 82.5 maf is the 75 maf required by law plus 7.5 maf for Mexico.) However, there is a major problem the statistics do not address.

Los Angeles is currently using Arizona's share of the Colorado River water. That water will have to be shifted back to Arizona upon completion of the Central Arizona Project (CAP), a 2 billion dollar Bureau of Reclamation effort designed primarily to deliver Colorado River water to the burgeoning cities of Phoenix and Tucson. Theoretically that would leave Los Angeles high and dry. In reality, that would never be allowed to happen, but it points out the seriousness of the dilemma facing water supply planners in the Southwest. There are alternative water supply sources that can be explored, but the alternatives are not promising.

*An acre-foot of water is one acre of land covered one foot deep in water.

SYNFUELS AND CATCH-22

Finally, there is the problem of the Upper Colorado River Basin. The northern portion, where Colorado, Utah, and Wyoming meet, is also the area in which the new synthetic fuels industry is likely to grow the fastest. Vast deposits of shale in that region hold an estimated 1.8 trillion barrels of oil, roughly sixty times the nation's proven reserves of liquid petroleum.*

Serious environmental problems are brought about by the production of oil from shale, however. One is that the crushed residue remaining after the mining process has a larger volume than the extracted shale. The rubble must be disposed of in an environmentally acceptable manner. The second problem is that the process of converting shale to oil requires a great deal of water—about two barrels of water for each barrel of oil produced. The shale must be heated to temperatures as high as 900°F to break it down into oil and gas, and water is needed for cooling purposes—as well as to flush impurities out of the shale. It is estimated that the shale industry could swallow almost 260,000 acre-feet of water annually.

Other energy development efforts in the Upper Basin will also demand large amounts of water. Coal-fired electrical generation plants may need as much as 475,000 acre-feet of water per year, largely for cooling functions. Coal gasification, coal liquefaction, and coal slurry (a mixture of powdered coal and water pumped through pipelines) will claim additional Upper Basin moisture. The U.S. Department of the Interior estimates that all pending energy development in the Upper Basin could use up almost 875,000 acre-feet of water each year.[21]

Given all of the projected requirements, significant water shortages could haunt every Upper Basin state except Wyoming by the year 2000. Add a 1930s-type drought to the scene early next century, and you have to wonder whether the extensive energy development envisioned in the West should be allowed to proceed. Remember, we are not gaining anything in terms of the CO_2 problem by following the synfuels route, and remember also that CO_2-produced warming could be a big factor in encouraging a dust bowl-type drought. The whole thing becomes a classic Catch-22 situation: the synfuels de-

---

*In reality, oil shale is not shale, but a variety of limestone—marl—that is permeated with a solid fossil fuel called kerogen. The kerogen supplies the oil and gas extracted from the "shale."

manding already scarce water while at the same time contributing to CO_2-induced warming that may lead to drought and even greater water scarcity.

Let us imagine a ten-year drought period with an average annual flow at Lee Ferry of 11.8 maf, such as occurred during the 1930s. The active storage capacity available in the Upper Basin in September 1974 was 23.6 maf. Let us assume that that amount of water is stored at the onset of the ten-year drought. Over the course of the decade 82.5 maf must be sent to the Lower Basin. That leaves 59.1 maf available for Upper Basin consumption, or 5.9 maf per year for the ten-year period. And, projected needs by late this century are already in excess of 6 maf per year.[22]

That is not the worst that could happen. Dendrohydrologists have determined that there was a decade in the late 1500s when the average yearly discharge of the Colorado at Lee Ferry sputtered to 9.7 maf. Is it unreasonable to expect that this might recur? Perhaps not, considering the greenhouse threat brings promise of permanent change in our atmospheric circulation patterns. There is no reason to believe that after a 1930s-type circulation pattern is established, natural climatic trends will eventually shift the westerlies back to lower latitudes (as happened in the 1940s) and bring life-giving rains to the dessicated regions. Once CO_2-induced warming becomes firmly entrenched, drought might only intensify.

With that in mind, let us contemplate another drought scenario, this one with a ten-year average annual flow of just 9.7 maf. Let us further assume that hydrologists have had the foresight and luck to bring the Upper Basin storage to its full potential capacity of 31 maf before the drought set in. Again, a ten-year total of 82.5 maf must be delivered to the Lower Basin. That leaves the Upper Basin with 45.5 maf of water during the drought decade. That is about 4.6 maf per year, far short of the more than 6 maf that will be used by early next century.[23]

Are there ways to avoid this? The obvious answer is to substantially reduce development of synthetic fuels and energy sources dependent upon coal. That would mitigate the greenhouse threat, lower water demand, and lessen the possibility of a 1930s-type (or worse) drought.

SIXTY CLOUDLESS DAYS

Failing that, we could explore ways of augmenting western water supplies. For instance, water from the Columbia or Snake River Basins in the Pacific Northwest could be shunted to southern California to make up for the Colorado River water that will have to be diverted back to Arizona from California once the CAP is completed. But the Pacific Northwest has water problems of its own. And as Secretary of the Interior, Cecil D. Andrus, said in 1978, "I am opposed to any new plan that could result in the movement of water from one state to another. That is not the way to solve a water crisis."[24]

Cloud seeding might increase precipitation in certain areas, but legal problems and doubtful results make that alternative unattractive. After a five-year cloud seeding program in the San Juan Mountains of the Upper Colorado River Basin, the Bureau of Reclamation reported that there was no "significant added precipitation."[25] Besides, for cloud seeding to work, there have to be clouds present. In times of true drought, seedable clouds are rare. In a 1974 drought over eastern Nebraska, skies were virtually cloudless for nearly sixty days.

The desalination of sea water has also been considered as a method of increasing water supplies. But skyrocketing energy costs mean substantial escalation in the cost of desalinated water. Desalination on a large scale does not seem to hold an economical answer to water shortage.

After examining these and other alternatives, a 1977 National Academy of Sciences study reported that it appears "no significant . . . augmentation is available in the immediate future."[26]

By the turn of the century the quest for water in the West may take on all the urgency of our current search for alternative energy sources. In fact, the two pursuits are not separate. By setting an energy course away from fossil fuels and synthetics now, we can soften the eventual impact of the greenhouse effect and at the same time lower the demand for water in the West. If we persist in our advance down the fossil and synthetic path, we may well find prolonged and severe drought lurking at the end of it.

5 THE LONG ISLAND EXPRESS

A warming climate and a concomitant shift in the pattern of the westerlies, a shift back to 1930s-type configurations, has implications for other weather phenomena besides temperature and precipitation. Specifically, these are hurricanes and tornadoes. Hurricanes thrive on warm ocean waters; a warming earth (and its oceans) encourages the genesis of even more tropical storms and hurricanes. And as the westerlies shift northward in response to the warming, the tracks of the great storms are altered, too. Regions that had been relatively free from hurricane strikes become more vulnerable. The new patterns in the westerlies, also foster new patterns in tornado threat. Tornadoes spawn in response to energy provided by warmth, moisture, and jet stream winds. And because these factors change with changing climate, so do the areas of maximum tornado threat.

HURRICANES

Hurricane and tropical storm activity in the Atlantic, the Caribbean, and the Gulf of Mexico reached a modern maximum in the 1930s. Twenty-one tropical cyclones blossomed in 1933, and seventeen in 1936. The current average (1946–1975) is nine, and the most active year in the last two decades was 1969, when there were thirteen storms. Figure 5–1 shows the trend graphically from 1885 through 1978.

In 1938 it had been almost seventy years since New England had been ripped by a major hurricane, and the damage had been confined to a relatively small area of southeastern New England.[1] So the fact that on September 21 there was a severe hurricane off Cape Hatteras, North Carolina, was met with relative indifference by both the public and the U.S. Weather Bureau. Everyone knew that hurricanes swerved out to sea before they could slam into New England.

By nightfall close to 700 people were dead in New England and on Long Island, New York. Property damage was awesome: 387 million dollarsworth at depressed 1938 prices.[2] New London, Connecticut, lay in ruins; Providence, Rhode Island, was under 13 feet of water; and Milton, Massachusetts (near Boston) had been blasted by winds of up to 186 mph. The Great New England Hurricane had thundered northward without swerving. It had raced from Cape Hatteras to Long Island at speeds of up to 70 mph, and later became known as the "Long Island Express."[3]

The retreat of the westerlies to the north during the 1930s not only went hand-in-hand with increased tropical cyclone development, but it sometimes—as in the case of the legendary 1938 storm—prevented the tropical invaders from following their normal *recurvature path*: moving toward the U.S. coastline from the south or

Figure 5–1. The annual variation in the number of tropical storms and hurricanes in the Atlantic, Caribbean, and Gulf of Mexico, 1885–1978. Activity reached a maximum during the 1930s. Twenty-one tropical cyclones blossomed in 1933, and seventeen in 1936. The modern average is nine per year. (Updated from *Weather Watch*, Walker and Company, copyright 1979 by Harold W. Bernard, Jr.)

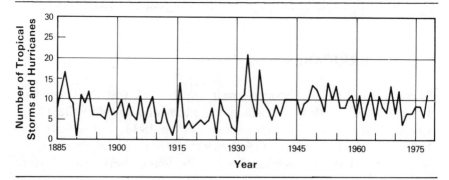

southeast (the tropical Atlantic Ocean), brushing, or in some cases entering, the Carolinas or Florida, then "recurving" to sweep eastward back across the Atlantic. Other eastern seaboard areas besides New England fell victim to changed hurricane tracks in the 1930s.

Several hurricanes swirled into the Cape Hatteras and Norfolk, Virginia, areas directly from the southeast during the 1930s, on tracks not since duplicated. The first of the storms struck on August 23, 1933, with the eye sweeping directly over Hatteras, then up the James River in Virginia. Winds reached 70 mph in Norfolk, and ten-foot tides surged through both Norfolk and Hampton, Virginia. High water marks hit eleven feet in the upper reaches of the Potomac River. Seashore resorts sustained major damage, with the final toll for the destruction reaching 27 million dollars. Fortunately, the death count was relatively low: eighteen.

Three weeks later a second tropical storm ripped into Cape Hatteras from the southeast, but hooked sharply away to the right before it could attack Hampton Roads. Tides ran about a foot lower than in the previous storm, with the Outer Banks of North Carolina catching the cyclone's full fury. Winds at Hatteras were clocked at 75 mph, and thirteen inches of rain fell. Twenty-one lives were lost, mostly in New Bern, North Carolina, where water two to four feet deep ran through the streets. The damage tally was 3 million dollars.

A third hurricane charged into Cape Hatteras from a southeasterly direction on September 18, 1936. But this powerful storm, with winds in excess of 135 mph over the open ocean, relaxed a bit as it brushed the Outer Banks and began to recurve. Still, tides of 5 to 6 feet above normal roiled through Hampton Roads.

The year 1933 was the most active tropical storm season on record, and Florida alone was pounded by four different storms that year. On September 3 and 4, Jupiter Inlet, on Florida's east coast, caught the most powerful blow. Winds of 125 mph ripped through the area, but residents found themselves in the calm eye of the cyclone for a full forty minutes.

Texas also felt the brunt of four distinct storms in 1933. On the same date that Jupiter Inlet was being battered, winds of 80 mph howled through Brownsville, Texas, and tides built to twelve to fifteen feet. Forty people died, and the damage count reached 17 million dollars.

The 1930s brought Florida more tropical storms and hurricanes than any other decade this century. An average of almost two storms per year swept into the state. The most dangerous one still holds the Western Hemisphere record for lowest sea level pressure.

Matecumbe Key, during the Labor Day storm of 1935, recorded an aneroid barometer reading of 26.35 inches. This is lower than the aircraft dropsonde reading of 26.73 inches, which was obtained just before hurricane Camille smashed into Mississippi in August 1969.[4] The 1935 monster swept the Keys with wind blasts in excess of 200 mph and storm tides of fifteen to twenty feet. A World War I veterans camp was inundated, and the railroad over the Keys washed away. Dollar damage from the storm was relatively minor, 6 million, but the death toll was 408.[5]

The 1930s have a message for us then. If we persist in our fossil fuel folly, then we should be prepared to deal with increasing numbers of, and perhaps increasing violence in, tropical storms and hurricanes early next century. It is a message to which we should pay attention.

TORNADOES

Tornadoes, as well as hurricanes, are likely to be affected by the changing weather patterns caused by the greenhouse effect. But examining tornado statistics from the 1930s and comparing them to current figures is less meaningful than doing the same thing with hurricane numbers* because the reporting and detection of tornadoes have improved tremendously over the last few decades. Increasing population densities, public awareness, weather radars, modern communications, and programs of the National Weather Service have caused the number of reported tornadoes to take a big jump: from 1685 in the 1930s to 7721 in the period 1970−78.[6]

On the other hand—thanks to a sophisticated tornado forecasting and warning effort by the National Severe Storms Forecast Center in Kansas City, Missouri—the death rate from twisters has shown a marked decrease, despite the ever growing U.S. population. During

*Tornadoes and hurricanes are often confused by the general public. Tornadoes are black funnels that dip down out of thunderstorm clouds. Tornadoes typically have a diameter of several hundred yards and a life span measured in minutes. Hurricanes are large storms born over tropical oceans. They may have a diameter of a few hundred miles and live for a week or more. The term "cyclone," as used by meteorologists, may apply to a hurricane, a typical wintertime storm, or a small-scale storm associated with a tornado, but not to a tornado itself. "Twister" and "funnel" are commonly used synonyms for tornado.

the 1930s, tornadoes took 1944 lives, while from 1970 through 1978 the death toll was 903.[7]

Still, there is a method of getting a pretty good idea of where tornado activity has been concentrated at various times in recent decades. This is done by studying tornado death anomalies, an approach taken by a group of meteorologists led by Dr. T. Theodore Fujita of the University of Chicago.[8] By finding areas of the country where the tornado death rate has been significantly above that which would have been normally expected (an anomaly is a departure from normal), the group was able to indirectly determine regions of maximum tornado threat.

One might expect that as the westerlies shift to somewhat higher latitudes, tornado action would follow suit. But this was not the case during the 1930s. There were several violent funnels that tracked across southern Minnesota, Wisconsin, and Michigan during the 1930s,[9] but it was the southeastern U.S. that became "tornado alley."

Early in the period (1926–1933 in the University of Chicago study), Wisconsin experienced slightly more than normal tornado fatalities, but maximum death rates—more than twice the average—were found in Missouri, Arkansas, western Tennessee, northern Mississippi, and northern Alabama.

Throughout the bulk of the 1930s (1934–1940 in the Chicago study), the largest anomalies were found along a band from Louisiana across the Gulf Coast states through the Carolinas. (A small anomaly was centered in southern Minnesota.) Georgia suffered the largest jump in deaths, with a fourfold increase in fatalities being reported.

On April 6, 1936, a swarm of twisters took 203 lives and wiped out the business district of Gainesville, Georgia. A day earlier the same weather system had unleashed a violent funnel near Tupelo, Mississippi. That storm left 216 people dead, 700 injured.

Other significant storms of the late 1930s ripped through central and northern Alabama. A tornado outbreak on March 21, 1932, devastated rural areas of many counties, wiping out 268 lives and injuring 1874. Louisiana was struck severely on May 1, 1933. Five hundred homes were damaged in Minden, in the northwest part of the state. Twenty-three people were killed, 400 injured.

A twister dipped down on Greensboro, North Carolina, on April 2, 1936, took 14 lives, and left 148 injured. In South Carolina, Charleston was battered by a tornado swarm on September 29, 1938. Dead: 32; injured: 150.[10]

Maximum tornado activity was not found farther north in the 1930s probably because of the great drought and attendant weather patterns that prevailed over the Midwest at that time. The lack of precipitation implies a relative lack of thunderstorm activity, and tornadoes must develop from thunderstorms. Thus, the twisters vented their anger on the deep South, where warmth and moisture — part of the tornado recipe — were more readily found.

Again, the message of the 1930s, the message of the greenhouse threat, is clear.

6 CLIMATIC STRESS

The westerlies reached their maximum northward extent in the early 1930s. The attendant climatic warmth of the time is reflected in the greater number of state records for high temperatures and dryness set during the 1930s than during any other decade since the 1870s. See Table 6−1.

After the early 1930s the westerlies began to display a greater frequency of looping (or blocking) patterns and we went into what Dr. Willett of MIT terms a period of *climatic stress*, which lasted through the 1950s.[1] When a blocking pattern is established, some regions become persistently hot and dry, others cold or wet. Table 6−1 shows that not only were the greatest number of state records for high temperatures set in the 1930s, so were the greatest number of state record lows.

The high latitude blocking configurations of the 1930s fostered persistent and anomalous weather patterns. Wind flows became "locked in" for extended periods at various times in different parts of the country. It was an era of more than record-shattering heat waves and drought. It was a decade of severe cold snaps and death-dealing floods, as well.

Table 6−1.

State[a] Record Maximum Temperature		State[a] Record Minimum Temperature	
Decade	Number of State Records	Decade	Number of State Records
1870−1879	0	1870−1879	0
1880−1889	1	1880−1889	1
1890−1899	2	1890−1899	7
1900−1909	2	1900−1909	6
1910−1919	5	1910−1919	5
1920−1929	3	1920−1929	3
1930−1939	27	1930−1939	11
1940−1949	1	1940−1949	4
1950−1959	6	1950−1959	5
1960−1969	1	1960−1969	6
1970−1978	3	1970−1978	3

State[a] Record Maximum Annual Rainfall		State[a] Record Minimum Annual Rainfall	
Decade	Number of State Records	Decade	Number of State Records
1870−1879	2	1870−1879	0
1880−1889	5	1880−1889	0
1890−1899	2	1890−1899	0
1900−1909	3	1900−1909	0
1910−1919	1	1910−1919	1
1920−1929	1	1920−1929	1
1930−1939	4	1930−1939	19
1940−1949	9	1940−1949	7
1950−1959	11	1950−1959	11
1960−1969	7	1960−1969	11
1970−1978	1	1970−1978	0

Other records were set in 1840, 1845   Another record was set in 1826
1851, 1853, and 1869

[a]Includes Washington, D.C.

HEAT WAVES OF THE 1930s

The midwestern heat began in the summer of 1930 and culminated in the summer of 1936. Between 1930 and 1936 nearly 15,000 people were killed by the sweltering temperatures. In a typical year in the United States about 175 people die from excessive summer heat and sunshine. In 1936 alone 4768 lives were taken by soaring temperatures and unrelenting sun.[2]

The first of the great heat waves appeared in the midsection of the nation in July 1930. Monthly temperatures averaged 6°F above normal in South Dakota and Nebraska, and 4°F greater than the mean in Arkansas and Tennessee. Nashville hit 104° four times during the month. The mercury soared to all-time state record levels in Tennessee, as well as in Kentucky and Mississippi. Perryville, Tennessee, tagged 113°F; Greensburg, Kentucky, registered 114°F; and Holly Springs, Mississippi, sizzled in 115°F heat. Readings reached 108° in Florida, but even hotter weather was to come a year later. In the East, Delaware tallied a state record 110°F, and Washington, D.C., hit a record-breaking 106°F.

The heat continued into August. August 3 became the hottest single day in Iowa history. The average high temperature throughout the state that broiling Sunday was 106.4°F. It was but a harbinger of things to come.

Record hot spells continued in 1931. In April Pahala, on the island of Hawaii, established the Hawaiian mark for heat with a reading of 100°F. Abnormally high temperatures in late June and early July in the Midwest killed thousands of field horses in Iowa. A June record in Minnesota was established when the temperature at Canby climbed to 110°F. Monticello, Florida, racked up that state's highest reading ever—109°F—on the same date, June 29th. On the West Coast, residents of San Diego were living through the warmest summer in local history.

By the time the year was done, figures showed 1931 to have been one of the warmest years on record. Every section of the country had had above normal temperatures. Minnesota had averaged 7.5°F above the mean, southern California, 4°F warmer than normal, and Washington, D.C., 3.5°F. In Montana it had been almost a degree warmer and an inch of precipitation drier than any previous year on the books.

The intense heat waves relented a bit during the next two years, 1932 and 1933. However, the winter of 1931–32 was the mildest ever around the Great Lakes and in parts of the Northeast; in Baltimore it was the warmest winter in 115 years. June heat baked parts of Oregon in 1932, and Blitzen set a monthly mark for the state with a 113°F high. The following June was a sizzler in the Midwest. In Iowa it was 2.1°F hotter than any previous June.

The oppressiveness returned in earnest in 1934. In Kansas and Iowa the summer was even hotter than in 1930, and was described as the worst crop season in history; the corn was virtually wiped out. Keokuk, Iowa, tallied a 118°F reading, and Gallipolis, Ohio, recorded 113°F. Both are state records for heat. Other all time state marks were established that summer in Idaho (Orofino, 118°F) and New Mexico (Orogrande, 116°F). In Cincinnati, Ohio, it was the hottest summer ever, with the thermometer topping off at 109° in July.

The heat backed off again in 1935, but, in a figurative sense, it was the calm before the storm. The heat during the torrid summer of 1936 was to become legendary.

In July and August of 1936, the heat records of 1934, which had topped those of 1930, were themselves exceeded. For example, July-August in Kansas averaged 80.5°F in 1930, 85.4°F in 1934, and 85.5°F in 1936. In Iowa the comparative figures were 76.1°F in 1930, 76.5°F in 1934, and 81.3°F in 1936! Mr. C.D. Reed, head of the Iowa Weather and Crop Bureau, made an extensive study of past heat conditions in his state. He drew these conclusions about July 1936: "Comparing the mean temperatures at individual stations in all Julys back to the beginning of records in 1819 with the July 1936 mean temperature in these localities, there is ample margin in favor of 1936 to take care of all possible differences due to location and exposure of instruments and methods of observation, and still leave July 1936 well in the lead as the hottest July in 117 years."[3]

The prostrating Iowa heat reached its pinnacle on the afternoon of July 14. The average maximum for 113 stations throughout the state was an astounding 108.7°F, topping the hot August Sunday of 1930 by better than 2°F.

Kansas City, Missouri, the largest city that borders on the plains, also put numerous records on the books that sweltering summer. The mercury soared to the 100°F mark or higher on fifty-three days. The first time came on June 15, with the peak on August 14 at

113°F, the city's hottest ever. The July–August period of 1936 was not only the steamiest on record in Kansas City, but at other cities in the Midwest and Great Plains as well: Cincinnati, Ohio; Louisville, Kentucky; St. Louis, Missouri; Des Moines, Iowa; Minneapolis, Minnesota; Bismarck, North Dakota; Pierre, South Dakota; Omaha, Nebraska; North Platte, Nebraska; and Oklahoma City, Oklahoma.

State records of 121°F were established in North Dakota and Kansas, and of 120°F in South Dakota, Texas, Arkansas, and Oklahoma. These are the only cases of readings of 120°F or higher ever being reached outside of the southwestern desert triangle of California, Nevada, and Arizona. There was one doubtful exception in Oklahoma in 1943 when a temperature of 120°F was reported, but no other station on the date in question came within 8°F of 120°F, and the figure is thus suspect. Elsewhere in the Midwest, Minden, Nebraska soared to 118°F during the 1936 hot spell, while locations in Minnesota and Wisconsin hit 114°F, all state records.

Figure 6–1. The average temperature deviations over the United States during the record hot month of July 1936. Maximum readings reached 120°F or more in North Dakota, South Dakota, Kansas, Oklahoma, Texas, and Arkansas. This was the only instance of temperatures of that level being reached outside of the southwestern desert triangle of California, Nevada, and Arizona.

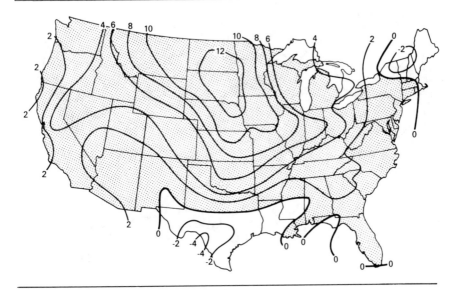

The superheat moved eastward and southward that summer, too, although not for extended invasions. Still, the hottest readings ever were recorded in Indiana (Collegeville, 116°F), Louisiana (Plain Dealing, 114°F), Maryland (Cumberland and Frederick, 109°F), Michigan (Mio, 112°F), New Jersey (Runyon, 110°F), Pennsylvania (Phoenixville, 111°F), and West Virginia (Martinsburg, 112°F).[4]

Figure 6-1 shows the average temperature deviations over the United States during that great hot month of 1936. For comparison, Figure 6-2 graphs the temperature deviations during a more recent hot month, July 1974. The heat anomalies of 1974 were just as intense as those of 1936, but were confined to a much smaller region. And, unlike July 1936, July 1974 was followed by an August much cooler than normal throughout the Midwest.

By the late 1930s the monster heat waves began to disappear, but not before leaving a state record of 117°F in Medicine Lake, Montana. Heat equal to the 1930s has not since returned to the United

Figure 6-2. The average temperature deviations over the United States during a modern hot month, July 1974. The heat anomalies of 1974 were just as extreme as those of July 1936, but were confined to a much smaller region. And, unlike July 1936, July 1974 was followed by an August much cooler than normal throughout the Midwest.

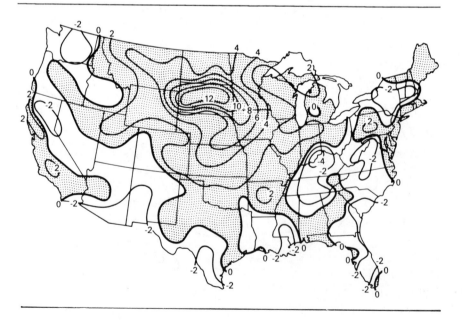

States, but as we inexorably alter our climate with CO_2, there are few reasons to believe that it will not. It could be one of the first actualizations of the greenhouse threat.

THE DANGERS OF HEAT

Extreme heat extracts a physical toll on us in a couple of ways. First of all, it puts a strain on our hearts. In hot weather our hearts work harder, pumping more blood to the tiny capillaries in the upper layers of our skin. There the blood—and thus the body—is cooled by giving off excess heat to the (relatively cooler) atmosphere. When the air temperature exceeds that of our bodies—normal body temperature is around 98.6°F—our bodies cannot shed heat through the circulatory system (if the environmental temperature is warmer than the temperature of our blood, there is no cooling effect). At this point, the only way we can lose heat is by sweating. The evaporation of the sweat cools the skin, and consequently the blood. But there are thermal limits to what our bodies can take, and when they are exceeded by very much or for very long, death threatens.[5]

A second way that hot weather affects us is through a change in metabolism. Metabolism refers to the chemical processes in our bodies that convert food and oxygen into tissue and energy. A by-product of metabolism is heat. When the outside air temperature rises to a point where getting rid of body heat becomes a problem, our metabolic rates slow down to curtail heat production. When metabolism becomes less active, so do our other bodily functions, including the capacity to fight off infections. That may explain why summer colds are sometimes so tenacious. One study of the records of acute appendicitis found that the disease struck twice as often during summer heat waves as during the winter.[6]

Another danger associated with heat waves is the pollution build-up that often accompanies them. The stagnant air masses that frequently settle in with stifling heat allow particulates, sulfur dioxide, carbon monoxide, ozone, and so forth to accumulate in particular areas. The resultant effects on our respiratory systems can be fatal. This is especially true in large cities where automobiles and heavy industries concentrate, and where the concrete and steel canyons among the skyscrapers deaden air circulation.

Typically, those individuals most influenced by heat and air pollution are the elderly and the poor—those most likely to reside in inner cities, those most likely to be in poor health to begin with, and those most likely to be unable to afford air conditioning.

Even at that, one has to wonder whether any air conditioners would be operating early next century under the onslaught of a 1930s-type heat regime. New construction by electric utilities is now met with frequent opposition, even in the face of an ever increasing demand for electricity. Unless this changes within the next decade or two, many utilities could be strained beyond capacity in their efforts to supply electricity during prolonged and severe hot weather. Instead of fresh air we may get brownouts, or even blackouts.

FORTY-ONE DAYS BELOW ZERO

The climatic stress of the 1930s produced not only hot spells, but some memorable cold waves, as well. The first of the frigid outbreaks dominated the West in January 1930. In Oregon it brought the most persistent January cold since the late-1800s. In parts of the Willamette Valley the month was the coldest on record; and in eastern Oregon the mercury dived to −52°F, a state record that was broken just three years later. In Montana the low reading for the month hit −52°F, in Wyoming, −57°F, in Minnesota, −49°F, in Illinois, −35°F (a mark that still stands), and in Oklahoma, a record −27°F. It was also the chilliest month in Texas history with a departure from normal of −9.4°F.

In 1932 a plunge of cold air into California set the stage for two unusual "snowstorms." On January 15 two inches of snow were judged to have fallen in downtown Los Angeles at the Civic Center. (The official measurement of snow on the ground was one inch.) Snow covered the beaches at Santa Monica, and up to eighteen inches coated mountain passes. The only other time since 1877 that measurable snow has fallen on Los Angeles was in January 1949 when less than half an inch sifted down.

The other anomalous California snow event in 1932 took place in San Francisco. On December 11 almost an inch of snow fell on the downtown area. It was the greatest snow since 1887, and until January 1962. That snowy 1932 day was also the coldest in San Francisco history, with a high of 35°F and a low of 27°F. At the

airport south of the city the mercury sank to an all-time record minimum of 20°F.

In February 1933 bitter cold once again invaded the West. State
records were stamped into the archives in Oregon (Seneca, −54°F),
Wyoming (Yellowstone Park, −66°F—the only colder temperature
ever observed in the lower forty-eight states was −70°F at Rogers
Pass, Montana in January 1954), and Texas (Seminole, −23°F). In
North Dakota, the coldest blizzard known swept the state. On February 8 Bismarck had a high of −17°F, a low of −35°F, and wind
gusts to 34 mph. Windchill factors were in the 70 to 80° below zero
range. The thermometer in Salt Lake City, Utah, sank to the lowest
level ever (−30°F) and the winter went on to become the coldest on
record. And in Phoenix, Arizona, residents watched in amazement
as an inch of January snow coated the desert. That has happened
only twice in Phoenix history—the other occurrence was to come
in 1937.

As 1933 ended, the arctic air shifted to the East. A late December
cold wave sent the mercury tumbling in New York (−47°F), New
Hampshire (−44°F), and Vermont (−50°F, a state record). The stage
was set for the numbing cold of February 1934.

On February 9, 1934, a bone-chilling −52°F settled over Still-
water Reservoir, New York, a record that was not equalled until
February 1979, when the thermometer again registered the same
nadir, this time at Old Forge. February 1934 saw Vanderbilt, Michigan, plunge to −51°F, a state record that remains. The month was
also the coldest in history at a number of major northeastern cities:
Boston, New York, and Buffalo. And it was the most frigid month
on Philadelphia records until January 1977.

That bitter 1934 February established record lows at Boston
(−18°F), Providence (−17°F), New York (−15°F), Buffalo (−21°F),
Philadelphia (−11°F), and in Michigan at Sault Saint Marie (−37°F).

As a point of modern comparison, February 1979 came very close
to duplicating the pattern of temperature departures from normal of
February 1934 in the northeastern United States. However, temperatures during the last week of February 1979 moderated; the persis-
tence of the 1934 cold was not matched.

The winter of 1934−35 was more moderate. In the fall of 1935
northern California experienced its earliest autumn freeze on record:
November 3 at Eureka, and November 4 at Sacramento. Then the
frigid air returned with a vengeance.

On November 30, 1935, the mercury edged below 32°F in Langdon, North Dakota, and did not rise above freezing until 92 days had passed, on February 29, 1936. For forty-one consecutive days, from January 11 to February 20, the temperature failed to rise above zero. It was the coldest extended period known in the United States, an ironic start for a year that was to bring the hottest summer on record to the same region. This was climatic stress!

February 1936 was the coldest month ever in U.S. history. In north central Montana temperatures averaged 26°F below normal. It was the coldest month known in Bismarck, North Dakota; Pierre, South Dakota; and Omaha, Nebraska. In Langdon, the mean temperature for the entire winter was −8.4°F. It was also the coldest winter on record in Des Moines, Iowa; Kansas City, Missouri; and Minneapolis, Minnesota (which suffered through thirty-six straight days with lows below zero); as well as in Pierre and Omaha.

New state marks for extreme cold were set in North Dakota (Parshall, −60°F), and South Dakota (McIntosh, −58°F), and local tallies were shattered in Bismarck (−45°F), Lander, Wyoming (−40°F), and Denver, Colorado (−30°F). Even the numbing cold of January 1979 in the northern Rockies and northern plains could not equal the spectacularly frigid February of 1936.

In the autumn of 1936, Denver was buried by a record early snow of 21.3 inches on 27 and 28 September. The stage was set for another wicked winter in the West.

And wicked it was. January 1937 established records for monthly cold all over the West: Helena, Montana; Lander, Wyoming; Denver, Colorado; Phoenix and Yuma, Arizona; and Sacramento, California. In Yuma the mercury crept down to a record minimum of 22°F; and for the second time within four years, an inch of snow fell on Phoenix, an event that has not been repeated since. Severe freezes plagued the California citrus crop, but the trees survived. Boca, in the northern California Sierras, recorded state record cold of −45°F; and San Jacinto, Nevada, set a record of −50°F.

While cold and snow covered the West, rain and warmth blanketed the East. Record floods surged through the Ohio and mid-Mississippi Rivers, and unusual warmth wiped out the winter sports season in New England. In Vermont only a little over a foot of snow fell in January and February, and monthly temperatures in January averaged almost 9°F above normal. In Boston, residents needed brooms, not shovels, to rid the area of the paltry nine inches of snow that came down that winter. Normal Boston snowfall is about forty-four inches per winter.

The anomalous 1930s cold spells closed out in 1938 with a parting shot at the Northeast. An early winter cold wave sent temperatures diving to below zero in late November—the earliest ever in Albany, New York, and Burlington, Vermont.[7]

DROWNINGS IN THE DUST BOWL

The climatic stress of the 1930s brought about not only great seasonal temperature contrasts, but contrasts in precipitation regimes as well. While the Midwest dried up under the onslaught of a great drought, record rains and floods decimated other regions. But even the Great Plains were subject to occasional intrusions of significant amounts of moisture. Before things really fell apart, agriculturally speaking, copious August rains in 1932 produced a record corn crop in Iowa. In 1935, excessive precipitation in the Republican and Kansas River Basins led to extreme floods in Colorado, Kansas, and Nebraska. Railroads were washed out for three weeks, and 110 people drowned, a true irony of the dust bowl days.

Floods in southern California in the 1930s took at least 146 lives. The worst of the washouts came in February 1938 when a series of storms dumped up to thirty inches of rain on the San Bernadino and San Gabriel Mountains. Huge floods roiled down the San Gabriel, Santa Ana, and Mohave Rivers. The death toll reached eighty-seven, and the property damage tally over 78 million dollars.

Just to the south of the heart of the dust bowl, inadequate precipitation was not a problem during the 1930s. Texas generally had greater than normal annual precipitation, and occasionally had to battle the results of too much rainfall. Floods came along fairly regularly in the mid- and late 1930s. One of the worst covered the state in September 1936. Torrential downpours kicked off widespread flooding, and a twenty-five-inch cloudburst at San Angelo washed away 300 homes.

Great precipitation contrasts were noted in the Pacific Northwest during the climatic stress years. In Washington state, the wettest year on record followed by just one year the driest ever. At Wahluke, in eastern Washington, a total of 2.61 inches of precipitation in 1930 was just barely enough to kick up dust. But in 1931, 184.56 inches inundated the Wynoochee and Oxbow areas of the Olympic Peninsula.

In Oregon, the greatest annual and least annual precipitation totals were separated by just two years. In 1937, Valsetz in the Coast Range was washed by 168.88 inches of rain and melted snow, while in 1939, the Warmsprings Reservoir, in the lee of the Cascades, caught only 3.33 inches.

In the eastern third of the nation, the story was not so much diversity in precipitation as too much precipitation.

March 1936 brought great floods to all of New England. Alternate periods of rain and thaw early in the month softened a vast blanket of ice and snow over the region. On the twelfth, heavy rains began to pelt down on the soggy mixture. David Ludlum, America's foremost weather historian, describes what happened after that:

> Massive snowslides occurred at high elevations, and rapid snowmelt began flooding valleys. More melting snow, thawing ice, and rainwater choked the normal channels of brooks, streams, and rivers and the waterways, which were soon ice jammed, were unable to handle the unprecedented volume. Dams gave way and bridges floated off or were smashed by debris; mills and factories were undermined, battered, and demolished; and most means of communications—highways, railroads, and wires—were severed. By the evening of the twelfth it was obvious over all New England that a major spring freshet was under way, but no one realized the magnitude of the mighty force that was moving toward the sea.[8]

The flood was especially severe in the Merrimack Valley of New Hampshire and Massachusetts. Water eighteen to twenty feet deep rushed through the main street of Hooksett, New Hampshire, and the Amoskeag Mills suffered tremendous damage. A flood marker at the Mills indicated the surge of water to be 13.5 feet higher than any previous crest. In the Massachusetts cities of Lawrence, Haverhill, and Lowell, mills and factories were totally inundated. In Maine, eighty-one bridges were swept away, as the three principal drainage basins, the Penobscot, the Kennebec, and the Androscoggin, all reached record flood levels.

Virtually the whole length of the Connecticut River reported unprecedented flood heights. At Hartford, where flood records had been maintained for more than 300 years, the 1936 crest topped the list by 8.6 feet at 37.6 feet. (Later, rain from the Great New England

Hurricane of 1938 brought the river to 35.4 feet, and the floods spawned by Hurricanes Connie and Diane in 1954 brought the level to 30.6 feet.) Much of downtown Hartford went under water in the 1936 flood when the Connecticut backed up into a small tributary stream flowing through the city.

Floods also swept through Pennsylvania in March 1936, and the Susquehanna River at Harrisburg rose 3.5 feet over its previous crest. (Hurricane Agnes in 1972 released floods on Harrisburg that topped the 1936 crest by four to six feet!) In all, 107 lives were lost throughout the Northeast, and the property damage totaled 270 million dollars. (By contrast, Agnes left over 3 *billion* dollars worth of destruction in her wake.)

Rampaging floods returned in 1937, but this time to the Ohio and Mississippi Rivers. A month of heavy rains throughout the Ohio Valley reached a climax in late January 1937. From the twenty-second to the twenty-sixth the Ohio was above flood stage along its entire length — 1,000 miles from Pittsburgh, Pennsylvania, to Cairo, Illinois. At Louisville, Kentucky, all previous flood stages were exceeded by ten to eleven feet; about two-thirds of the city's residential area, and almost all of its business district were awash. At Cincinnati, Ohio, the water surged to almost nine feet higher than any previous crest, and 10 percent of the city went under water.

In southern Illinois, about 90 percent of Gallatin County was covered by water. Nearly all of the 35,000 residents of Paducah, Kentucky, had to evacuate. Some river towns were completely abandoned, later to be rebuilt on higher ground.

South of Cairo, on the mid-Mississippi River, new high water marks were set for 350 miles. In all, the floods along the Ohio and Mississippi that January claimed 250 lives and did 470 million dollars-worth of property damage.[9] At their height they had inundated 12,700 square miles of land and 75,000 homes in twelve states.[10]

NOW IS THE TIME

It is apparent then, that climatic stress and a 1930s-type climate regime mean more than heat and dust and heightened tropical storm activity. They also mean paralyzing cold waves and destructive floods. They mean having to deal with climatic extremes at both ends of the weather spectrum. To be sure, drought presents the greatest danger, but our experiences with the frigid winters of the late 1970s showed us how much at the mercy of cold and snow our society remains.

Our ability to deal with climatic stress is suspect. Agricultural and economic planning are built around "normals," not extremes. The greenhouse threat promises not only to change those normals, but the extremes as well. We may well have difficulty in adjusting to the alterations in the normals. The struggle we might have in coping with the enhanced extremes could be worse.

Remember, in talking of a return to a 1930s-style climate, we are talking of only a brief stop along the way, the first stages of the greenhouse effect, the first problems we may have to face. If we persist in our pattern of fossil fuel consumption, the succeeding decades could bring a cascade of immense challenges—or immense disasters.

It would be far easier to meet those challenges now.

7 THE WEST COAST AND ALASKA—THE 1930s

The following five chapters take a closer look at the overall climatological conditions that prevailed across the U.S. (with the exception of Hawaii) during the 1930s. By comparing the averages of temperature and precipitation for the decade 1930 through 1939 with contemporary normals, we can perhaps get a feeling for what the initial stages of anthropogenic warming—the greenhouse effect—may be like in specific regions. Contemporary normals, as defined by the National Weather Service, are the averages of temperature, precipitation, snowfall, wind, and so on—that prevailed during the thirty-year period 1941–1970. After 1980 these "normals" will be updated, and will be for the thirty years 1951–1980.

The most anomalous meteorological events of the 1930s—drought, floods, cold waves, hot spells—have been discussed in previous chapters. The ensuing chapters will look more at mean climatic patterns on a month-by-month basis, and examine averages from the 1930s to see how they compare with current averages—the climatic conditions we are used to. The statistics from the 1930s ought to have a great deal of relevance to our climatic future in the 2010–2020 time frame.

For the purposes of the comparison I studied the weather records of thirty-seven different cities in the United States. The location of the cities, with the exception of Fairbanks, Alaska, is shown in Figure 7–1. The specific cities were chosen for a number of reasons.

Figure 7–1. The location of the cities whose weather records were studied as the basis for a comparison of the climatic averages of the 1930s with current normals. The climatic conditions of the 1930s may be indicative of what the initial stages of anthropogenic warming—the greenhouse effect—will be like.

First, an attempt was made to get well-spaced coverage across the country. Second, I tried to use only records from stations that had not moved their observing locations between 1930 and 1970. In many cases this was not possible, so I instead chose stations that had made only minor moves in location. In some cases this meant having to use records from stations outside major population centers, such as Rockford, Illinois (instead of Chicago), or Rome, Georgia (instead of Atlanta.)

In a few cases, no alternatives were available, and I had to use records from stations that had shifted their observing sites from city to rural airport settings. Such a shift can lead to substantial differences in air temperature, because of the urban "heat island" effect. Metropolitan areas can produce and retain a great deal of heat; on some clear, calm, and dry nights, temperatures in outlying regions can be as much as 10°F or 12°F lower than city readings.

A graphic example of this is shown by Columbia, Maryland, a town located northeast of Washington, D.C., and built from the ground up in the late 1960s. Within two years, the site of the town changed from rural to urban, and by 1970, with the population of Columbia at 7,000, average temperatures had risen 2°F over those of the surrounding countryside.[1]

A third criterion for choosing the cities was to find observational sites that had not only not moved, but that had also kept their thermometers at pretty much the same height above ground. Modern extreme temperature observations are taken four to six feet off the ground, but this practice was not strictly adhered to in the 1930s. On a bright, sunny afternoon, readings taken from a thermometer exposed 200 feet above the surface can run 1°F or more lower than observations recorded near the ground. On the other hand, on a clear, calm, cold night, temperatures at the 200-foot level might be substantially higher than those close to the surface. Unfortunately, consistency in thermometer exposure was rarely available over the period in which we are interested (1930 to 1970).

Still, a number of the cities I picked had good long-term records. With a little subjective correction for variations of instrument exposure and location, some meaningful, coherent patterns of temperature and precipitation differences between the 1930s and more recent times emerged. The patterns, for the reasons just discussed, may not be precise, but they capture the essence of the differences. Precipitation measurements, by the way, are much less affected by site moves than are temperature readings. Assuming that any station move is of only a few miles and does not result in significantly altered elevation or topography, precipitation averages at the old and new locations should be virtually the same.

THE WEST COAST

While much of the midsection of the nation was noticeably warmer in the 1930s than it is now, this is not the case along the Pacific Coast. Records from Salem, Oregon (Table 7-1), indicate that temperatures there during the 1930s were about the same as they are now, while data from San Francisco, California (Table 7-2), and Los Angeles, California (Table 7-3), indicate the 1930s may actually have been slightly cooler along the coast of the Golden State.

In the Pacific Northwest the numbers from Salem—as well as those from Walla Walla, Washington (Table 8-1)—tell of more wintry Februarys through the 1930s, with colder mean temperatures and

Table 7–1.   Monthly Averages, Salem, Oregon.

Months	Temperature (Degrees F)		Precipitation[a] (Inches)		Snow (Inches)	
	1930–1939	1941–1970[b]	1930–1939	1941–1970[b]	1930–1939	1941–1970[b]
January	38.7	38.8	6.37	6.90	4.0	4.4
February	41.4	42.9	5.05	4.79	2.8	0.7
March	46.7	45.2	4.79	4.33	0.3	0.9
April	51.8	49.8	2.69	2.29		
May	56.0	55.7	1.87	2.09		
June	61.5	61.2	1.36	1.39		
July	65.7	66.6	0.28	0.35		
August	66.0	66.1	0.35	0.57		
September	61.5	61.9	1.24	1.46		
October	54.0	53.2	3.14	3.98	0.5	
November	44.6	45.2	4.92	6.08	*	0.2
December	41.3	40.9	9.04	6.85	*	1.3
Annual	52.4	52.3	41.10	41.08	7.6	7.5

* Less than 0.05 inches.

[a] Includes melted snow.

[b] Current normals.

more snowfall than currently. However, most of the February increase in average snowfall at Salem during the 1930s was due to just one massive storm—in February 1937—that dumped over twenty-five inches on Oregon's capital city. That was enough to make the month the second snowiest on record there. The coldest month on the books at Salem—by over 2°F—was January 1930, when the mercury averaged 27.4°F.

December 1932 brought a deep chill to central California as both San Francisco and Sacramento reached their all-time temperature nadirs—20°F on the bay, and 17°F in the valley at Sacramento. At the northern end of the Sacramento Valley, Red Bluff set its record temperature minimum in January 1937, also 17°F. But despite the occasional icy winter readings that prevailed in California during the 1930s, the most notable temperature differences occurred in the summer with both San Francisco and Los Angeles averaging roughly a degree cooler than current means.

Table 7-2. Monthly Averages, San Francisco, California.

Months	Temperature (Degrees F)		Precipitation[a] (Inches)	
	1930–1939	1941–1970[b]	1930–1939	1941–1970[b]
January	47.5	48.4	3.93	4.37
February	50.6	51.1	3.67	3.04
March	53.2	52.6	2.84	2.54
April	54.7	54.6	0.92	1.59
May	57.4	57.3	0.35	0.41
June	59.8	60.3	0.15	0.13
July	60.8	61.6	0.00	0.01
August	61.0	62.2	0.02	0.03
September	61.8	63.3	0.13	0.16
October	59.6	60.2	0.63	0.98
November	54.1	54.7	1.09	2.29
December	48.8	49.6	3.56	3.98
Annual	55.8	56.3	17.29	19.53

Note: Extreme thermometers at 28 to 41 feet above ground through 1930s.

[a] Includes melted snow.

[b] Current normals.

Precipitation averages from 1930 through 1939, as compared to modern normals, showed little change at Salem, a slight decrease at San Francisco, and an increase at Los Angeles. All three stations had markedly less rainfall in November then than they do now, while both Salem and Los Angeles experienced noticeably wetter Decembers.

Although precipitation during the 1930s was little different from currently at Salem, the depression decade brought the wettest month ever recorded there (December 1933 with 17.54 inches of "Oregon sunshine") and the wettest year (1937 with 63.50 inches of rain and melted snow.)

In southern California, despite the overall increase in rainfall at Los Angeles, there were some drought years, particularly in the San Joaquin Valley. Fresno and Bakersfield both tallied their driest twelve months ever from July 1933 through June 1934. Fresno measured only 4.43 inches of rain, and Bakersfield but 2.26 inches.

Table 7−3.   Monthly Averages, Los Angeles, California.

Month	Temperature (Degrees F) 1930–1939	Temperature (Degrees F) 1941–1970[b]	Precipitation[a] (Inches) 1930–1939	Precipitation[a] (Inches) 1941–1970[b]
January	56.7	56.7	3.41	3.00
February	57.5	58.1	3.94	2.77
March	60.2	59.2	2.33	2.19
April	62.4	61.7	0.91	1.27
May	63.9	64.7	0.21	0.13
June	67.0	68.0	0.12	0.03
July	71.7	73.2	0.00	0.00
August	72.7	74.1	0.02	0.04
September	70.8	72.7	0.62	0.17
October	67.9	68.4	0.43	0.27
November	64.3	62.7	0.79	2.02
December	59.5	58.1	3.82	2.16
Annual	64.5	64.8	16.60	14.05

Notes: Extreme thermometers at 159 feet above ground through 1930s; data from city/ Civic Center locations.

[a] Includes melted snow.

[b] Current normals.

As was noted in Chapter 5, the 1930s produced some unusually active tropical storm seasons in the Atlantic, the Caribbean, and the Gulf of Mexico. But there was also some surprising activity in the eastern Pacific Ocean off the west coast of Mexico. Tropical storms and hurricanes that form off the Mexican Pacific coast typically move westward into the open ocean, or, if they do journey northward, weaken significantly as they approach the United States. However, in 1939, an anomalous number of full-fledged tropical cyclones—five—crossed the coast of North America from Baja California northward.

Two storms passed less than 200 miles south of San Diego, while a third one, on September 25, blasted inland near Los Angeles. Unprecedented gales and heavy rains caused damage in excess of 2 million dollars and took forty-five lives. Los Angeles was drenched with 5.5 inches of rain, while Mt. Wilson was deluged with 13 inches.[2]

It was not until 1976 that another tropical storm brought so much rainfall and destruction to southern California. In September 1976 Tropical Storm Kathleen charged up the Gulf of California and ripped into the lower Colorado Valley. The storm left a trail of flash floods, mudslides, and wind damage from the southern California deserts and western Arizona into Nevada. At least seven people were killed and damage mounted to 160 million dollars.[3]

In summary, the 1930s averaged a bit cooler—compared to modern normals—along much of the Pacific coast, while precipitation totals were unchanged from or slightly less than current means—except in southern California. There, the unique visitation from a full-blown tropical storm contributed to greater amounts of rainfall than normally occur now.

ALASKA

Data from Fairbanks, Alaska (Table 7–4), are presented more for curiosity reasons than for the purpose of trying to associate them with patterns of differences—between the 1930s and now—in the lower forty-eight states.

Fairbanks, throughout the 1930s, averaged colder, wetter, and snowier than now. Winters ran almost a degree colder than current winters, but summers showed an even greater departure from modern normals, averaging about 1.5°F cooler. The most notable temperature event of the decade was a winter one, however. Everything but the thermometer froze up in January 1934 when the mercury dived to −66°F, a record that still stands. (December 1961 came close: −62°F.) Another all-time mark was tallied in January 1937 when 65.6 inches of snow buried the town. (The greatest monthly amount since then was 54 inches in November 1970.)

Thanks to some very wet Augusts, annual precipitation totals at Fairbanks were a bit higher in the 1930s than now. August 1930 was the wettest month on record there: 6.88 inches. Other Alaskan precipitation records worthy of note during the 1930s were the 1.61 inches that fell at Barrow in 1935 (a state mark for least annual precipitation) and the 119. 48 inches that drenched Juneau in 1937. The closest rainfall in modern times was in 1961: 68.11 inches.

As a point of interest, New Orleans, Louisiana, with an average annual rainfall total of 56.77 inches is wetter than Juneau where the

Table 7-4.   Monthly Averages, Fairbanks, Alaska.

Months	Temperature (Degrees F) 1930–1939	1941–1970[b]	Precipitation[a] (Inches) 1930–1939	1941–1970[b]	Snow (Inches) 1930–1939	1941–1970[b]
January	−12.0	−11.9	1.29	0.60	17.0	10.5
February	−6.8	−2.5	0.51	0.53	9.2	9.3
March	7.5	9.5	0.47	0.48	7.7	7.7
April	29.5	28.9	0.18	0.33	2.1	4.2
May	45.7	47.3	0.59	0.65	0.5	0.7
June	57.7	59.0	1.21	1.42		
July	59.3	60.7	1.92	1.90		
August	53.6	55.4	3.11	2.19		
September	43.8	44.4	0.89	1.08	0.6	0.5
October	27.3	25.2	1.13	0.73	10.2	8.7
November	3.4	2.8	0.85	0.66	10.8	9.9
December	−8.6	−10.4	0.56	0.65	10.4	9.6
Annual	25.0	25.7	12.71	11.22	68.5	61.1

Source: Data from city location through 1930s.
[a] Includes melted snow.
[b] Current normals.

normal yearly precipitation adds up to 54.67 inches. It normally takes a bit longer at Juneau, however: 220 days, versus 114 days for New Orleans.

8 THE INTERMOUNTAIN WEST, THE ROCKIES, AND THE SOUTHWEST — THE 1930s

As we shift our study inland from the West Coast and examine 1930s weather records from the interior West, we can begin to see the first real manifestations of the warmth and dryness that characterized the 1930–1939 decade. As I have mentioned, the most intense heat and dust of that period occurred in the center of the United States, but droughty weather spilled over into areas west of the Rockies, too.

THE INTERMOUNTAIN WEST

Records from Walla Walla, Washington (Table 8–1), and Salt Lake City, Utah (Table 8–2), although indicating that the 1930s were little different in temperature from now, also tell us that the period 1930–1939 was much drier than the current one. At Walla Walla, average annual precipitation for the decade was 85 percent of the modern normal, while at Salt Lake City it was just 81 percent. Seven out of ten years at Walla Walla, and eight out of ten at Salt Lake City during the 1930s had yearly precipitation amounts less than the current average.

Other western cities besides Salt Lake City and Walla Walla suffered from the decade's dessication, of course. In Idaho, Lewiston's driest year on record was 1935 (8.40 inches) and Pocatello's was

Table 8–1. Monthly Averages, Walla Walla, Washington.

Month	Temperature (Degrees F)		Precipitation[a] (Inches)		Snow (Inches)	
	1930–1939	1941–1970[b]	1930–1939	1941–1970[b]	1930–1939	1941–1970[b]
January	33.7	33.4	1.55	2.07	7.2	7.6
February	36.8	40.2	1.51	1.40	6.4	2.5
March	47.0	45.6	1.88	1.37	1.4	1.4
April	54.8	52.8	1.20	1.43	0.1	
May	61.4	60.3	1.06	1.58		
June	68.1	67.2	0.95	1.18		
July	76.3	75.6	0.15	0.33		
August	74.4	73.6	0.10	0.45		
September	66.3	65.7	0.60	0.85		
October	55.6	54.4	1.23	1.49	0.1	
November	42.5	42.7	1.65	1.89	1.0	1.5
December	37.6	37.0	1.71	1.97	3.5	4.5
Annual	54.5	54.1	13.59	16.01	19.7	17.5

Source: Data from city location.
[a] Includes melted snow.
[b] Current normals.

1939 (6.43 inches). In the far Northwest, the driest year at Yakima, Washington, was 1930, when only 3.9 inches of precipitation fell; Burns, Oregon, measured only 6.87 inches from July 1938 through June 1939 to establish an all-time twelve-month low there.

In the western reaches of the intermountain area, observations from Winnemuca, Nevada (Table 8–3), indicate that annual precipitation amounts in the 1930s did not vary significantly from modern totals. Annual temperatures, however—the records tell us—were somewhat milder, but with notable seasonal differences. As was the case through all of the intermountain West, winters in the 1930s were colder than those of more modern times, while summers were hotter and drier.

In Salt Lake City, the winter of 1932–1933 was the most frigid on record with the mercury averaging only 19.5°F. However, despite the greater frequency of chilly winters in the 1930s, snowfall totals

Table 8-2. Monthly Averages, Salt Lake City, Utah.

Month	Temperature (Degrees F)		Precipitation[a] (Inches)		Snow (Inches)	
	1930–1939	1941–1970[b]	1930–1939	1941–1970[b]	1930–1939	1941–1970[b]
January	25.7	28.0	1.11	1.27	13.0	13.4
February	31.9	33.4	1.37	1.19	10.5	9.7
March	41.1	39.6	1.34	1.63	6.0	10.8
April	50.8	49.2	1.22	2.12	0.7	5.2
May	58.3	58.3	1.23	1.49	0.1	0.6
June	68.5	66.2	0.59	1.30		
July	77.1	76.7	0.56	0.70		
August	74.6	74.5	1.02	0.93		
September	64.9	64.8	0.60	0.68		0.1
October	52.6	52.4	1.12	1.16	0.4	0.8
November	38.3	39.1	1.03	1.31	5.8	6.2
December	29.9	30.3	1.16	1.39	10.7	12.4
Annual	51.1	51.0	12.35	15.17	47.2	59.2

[a] Includes melted snow.

[b] Current normals.

in Utah's capital averaged a bit less then than now. This was largely because of a reduction in March and April snowfalls there. Two factors probably contributed to those diminished spring snows. March and April mean temperatures through all of the intermountain West were a bit higher in the 1930s than currently; and the precipitation (rain and melted snow) total at Salt Lake City for those two months was only 68 percent of the current normal.

The colder winters that prevailed in the intermountain region in the 1930s were frequently offset by hot, dry summers. Salt Lake City's hottest month ever was July 1933 (81.3°F). That went along with the driest summer on the books there—June, July, and August of 1933 when less than a quarter-inch of rain fell. Farther west, Reno, Nevada, tallied its hottest month known in July 1931: 77.4°F. And the summer of 1931 went on to become the all-time sizzler in Reno's history.

Table 8–3. Monthly Averages, Winnemucca, Nevada.

Month	Temperature (Degrees F)		Precipitation[a] (Inches)		Snow (Inches)	
	1930–1939	1941–1970[b]	1930–1939	1941–1970[b]	1930–1939	1941–1970[b]
January	27.5	28.8	1.17	0.97	10.9	7.0
February	31.4	34.9	1.07	0.81	6.8	3.8
March	40.9	38.3	0.85	0.71	4.5	5.0
April	48.9	45.9	1.09	0.73	1.4	2.0
May	56.1	54.6	0.95	0.91	0.4	0.4
June	65.2	62.3	0.51	1.01		
July	73.9	72.1	0.28	0.23		
August	71.4	68.6	0.12	0.26		
September	61.3	59.9	0.33	0.28		
October	50.1	49.0	0.80	0.65	0.2	0.4
November	37.6	37.9	0.57	0.97	2.8	2.4
December	31.0	30.9	0.90	0.94	4.9	5.0
Annual	49.6	48.6	8.64	8.47	31.9	26.0

Source: Data from city location through 1930s.

[a] Includes melted snow.

[b] Current normals.

THE ROCKIES

Temperature and precipitation patterns of the 1930s—relative to current normals—were much the same in the Rocky Mountains area as they were in adjacent regions of the intermountain West. Data from Sheridan, Wyoming (Table 8–4), and Denver, Colorado (Table 8–5), tell us that superimposed on a very small annual warming in the 1930s—compared to now—were slightly colder winters, and hotter and drier summers. The month of June in particular, at both Sheridan and Denver, was noticeably hotter and drier during the 1930s than now. And summers as a whole produced only about 65 to 70 percent of currently normal precipitation for the three-month period.

Mean annual precipitation amounts were down significantly, too. Denver measured only 81 percent of its current average yearly pre-

Table 8−4.   Monthly Averages, Sheridan, Wyoming.

Month	Temperature (Degrees F) 1930−1939	Temperature (Degrees F) 1941−1970[b]	Precipitation[a] (Inches) 1930−1939	Precipitation[a] (Inches) 1941−1970[b]	Snow (Inches) 1930−1939	Snow (Inches) 1941−1970[b]
January	20.1	21.0	0.58	0.69	6.7	9.7
February	23.1	25.9	0.59	0.77	7.1	11.1
March	32.4	31.0	1.52	1.21	12.0	13.1
April	44.4	43.6	1.75	2.12	7.7	9.3
May	54.8	53.1	2.68	2.45	0.1	1.7
June	63.5	61.1	1.82	2.99		0.2
July	72.2	70.4	1.06	1.07		
August	68.7	69.2	0.71	0.95		
September	58.4	57.9	0.85	1.28	0.9	0.6
October	46.5	47.8	1.36	1.02	3.6	3.2
November	34.4	33.4	0.49	0.92	6.2	8.4
December	26.2	25.5	0.48	0.69	5.5	9.5
Annual	45.4	45.0	13.89	16.16	49.8	66.8

Source: Data from city location through 1930s, now airport at 170 to 190 feet higher.
[a] Includes melted snow.
[b] Current normals.

cipitation, while Sheridan got 86 percent. At both locations seven out of ten years in the 1930s had less precipitation than the modern normals. The parched 1930s brought record dryness to several parts of Colorado. In the central Colorado Rockies, Buena Vista caught just 1.69 inches of rain and melted snow in 1939 to establish a state record for lack of precipitation. In the same year Colorado Springs set its mark for dryness: 6.07 inches. Earlier in the decade, Pueblo, Colorado, near the heart of the dust bowl, had had its dustiest year on record: 1934, with but 5.78 inches of precipitation.

THE SOUTHWEST

Most of the southwest United States—Arizona and New Mexico—in contrast to much of the rest of the country, was damper during the 1930s than currently. At least that is the indication of the records

Table 8–5.   Monthly Averages, Denver, Colorado.

Month	Temperature (Degrees F) 1930–1939	1941–1970[b]	Precipitation[a] (Inches) 1930–1939	1941–1970[b]	Snow (Inches) 1930–1939	1941–1970[b]
January	28.8	29.9	0.33	0.61	5.5	8.6
February	32.3	32.8	0.76	0.67	10.2	8.1
March	38.4	37.0	1.06	1.21	10.6	13.5
April	48.2	47.5	1.91	1.93	10.7	9.2
May	56.7	57.0	2.36	2.64	1.6	1.9
June	68.0	66.0	0.88	1.93		
July	74.6	73.0	0.93	1.78		
August	72.2	71.6	1.39	1.29		
September	64.0	62.8	1.12	1.13	1.9	1.2
October	52.0	52.0	0.72	1.13	3.9	4.0
November	39.5	39.4	0.56	0.76	9.4	7.9
December	32.7	32.6	0.49	0.43	7.2	5.6
Annual	50.6	50.1	12.51	15.51	61.0	60.0

Note: The 1930–1939 temperature data were "normalized" to (or made compatible with) the current airport location by subtracting the mean monthly differences—as calculated by the U.S. Weather Bureau for the period 1921–1950—between the two locations.

[a] Includes melted snow.

[b] Current normals.

from Phoenix, Arizona (Table 8–6), and Albuquerque, New Mexico (Table 8–7). The Albuquerque observation site has moved around in the past fifty years, so the records there lack consistency of location. Still, there are some general trends worth noting. Albuquerque, throughout the 1930s, probably was somewhat cooler than now.

Although the shifts in the Albuquerque observation location may have affected precipitation records to some extent, general indications are that most winters of the 1930s there were a bit drier than modern ones, and the summers a bit wetter.

The biggest climatic difference between Albuquerque and Phoenix during the 1930s occurred in the summers. Phoenix summers were relatively hotter and drier than they are today, much like summers to the north (in the intermountain region) were, while Albuquerque

Table 8-6.   Monthly Averages, Phoenix, Arizona.

Month	Temperature (Degrees F)		Precipitation[a] (Inches)	
	1930–1939	1941–1970[b]	1930–1939	1941–1970[b]
January	51.2	51.2	0.78	0.71
February	55.9	55.1	1.36	0.60
March	62.1	59.7	0.65	0.76
April	70.4	67.7	0.22	0.32
May	77.5	76.3	0.16	0.14
June	86.7	84.6	0.09	0.12
July	92.7	91.2	0.61	0.75
August	90.5	89.1	0.77	1.22
September	85.1	83.8	1.17	0.69
October	73.0	72.2	0.28	0.46
November	60.6	59.8	0.66	0.46
December	54.0	52.5	0.84	0.82
Annual	71.6	70.3	7.59	7.05

Source: Data from city location through most of 1930s with extreme thermometers at 10 to 39 feet above ground.

[a] Includes melted snow.

[b] Current normals.

summers were cooler and moister than today. The hottest month in Phoenix's history occurred early in the decade. July 1931 averaged 95.2°F. (More recently, July 1977 averaged 95.0°F.) Overall the average annual temperature at Phoenix throughout the 1930s was higher than the current normal.

Phoenix, like Albuquerque, experienced greater mean annual precipitation through the 1930s compared to current times. The most notable similarity in the records of the two stations was the marked increase in September rainfall.

But, the most remarkable of all weather events in the Southwest during the 1930s was the snow in Phoenix, not once, but three times! January 1933 and January 1937 each brought an inch of snow to the normally mild winter desert. The one-inch totals are records for

Table 8–7.   Monthly Averages, Albuquerque, New Mexico.

Month	Temperature (Degrees F) 1930–1939	Temperature (Degrees F) 1941–1970[b]	Precipitation[a] (Inches) 1930–1939	Precipitation[a] (Inches) 1941–1970[b]	Snow (Inches) 1930–1939	Snow (Inches) 1941–1970[b]
January	32.8	35.2	0.33	0.30	1.9	1.9
February	39.1	40.0	0.31	0.39	1.0	1.7
March	45.8	45.8	0.33	0.47	0.8	1.8
April	54.7	55.8	0.56	0.48	0.4	0.3
May	62.7	65.3	0.77	0.53	0.2	
June	72.7	74.6	1.01	0.50		
July	76.9	78.7	1.55	1.39		
August	75.5	76.6	1.33	1.34		
September	68.0	70.1	1.32	0.77		
October	56.8	58.2	0.55	0.79		
November	42.7	44.5	0.56	0.29	0.8	1.0
December	35.6	36.2	0.27	0.52	1.0	2.8
Annual	55.3	56.8	8.89	7.77	6.1	9.5

Note: Location has shifted several times since 1930; now airport at 200 feet higher elevation.

[a] Includes melted snow.

[b] Current normals.

Phoenix; a February record was set in 1939 when half-an-inch sifted down. No measurable snow has fallen on Phoenix since.

In summary, residents of the interior West in the 1930s suffered through icier winters and hotter and drier summers than commonly occur now. On an average annual basis, temperatures were perhaps as much as a degree warmer in the 1930s from Arizona up into Nevada, elsewhere only very minor warming was apparent, except for New Mexico, where readings averaged somewhat lower. Relatively drier weather held sway most of the time over the inland West during the 1930s, but much of Arizona and New Mexico were a bit less parched then.

9 THE GREAT PLAINS AND TEXAS—THE 1930s

The Great Plains of the United States suffered more than any region during the sweltering summers and parched years of the dust bowl era. The states of Kansas, Nebraska, and Iowa were particularly hard hit. From 1930 through 1939 average annual precipitation totals in those states were 80 percent or less of modern normals.

The most devastating effects of the warmth and dryness of the 1930s have been discussed above, in Chapter 4. This chapter will look at the overall, general climate of the dust bowl days on the Great Plains and throughout Texas. The climatic extremes are "smoothed out" in this view, but the ten-year trends in warmth and dryness are still startling.

THE NORTHERN PLAINS

The weather records from Havre, Montana (Table 9—1), Bismarck, North Dakota (Table 9—2), Minneapolis–St. Paul, Minnesota (Table 9—3), Valentine, Nebraska (Table 9—4), and Lincoln, Nebraska (Table 9—5)—the northern plains—tell of slightly milder winters, and markedly hotter and drier summers in the 1930s, as compared to more modern times. On an annual basis temperatures ran about 1.5° to 2.0°F higher during 1930—1939 than currently.

Although there were some bitterly cold winters—such as the one of 1935–1936 (see Chapter 6)—in the Midwest during the 1930s, winters overall in the northern plains averaged a bit higher in temperature then than now. But, the biggest differences in mean monthly temperatures between the 1930s and now were in the summer months. Julys, in particular, were noticeably hotter, averaging 4°F or more warmer over much of South Dakota, Nebraska, and Iowa. The mean July temperature differences between the 1930s and now are diagrammed in Figure 9–1 for all of the United States (except Alaska and Hawaii.)

Individual Julys during the dust bowl decade had mean temperature departures considerably above 4°F, of course. For instance, at Lincoln, July 1936 averaged almost 10°F above the modern mean! An *average* difference of 4°F or more persisting for ten years is really quite remarkable when you consider that, at Lincoln—for example—the hottest single Julys since the 1930s have averaged only about 4° or 5°F above the current normal.

Figure 9–1. The mean July temperature differences between the 1930s and now. Julys of the 1930s averaged noticeably hotter—4°F or more—over much of South Dakota, Nebraska, and Iowa.

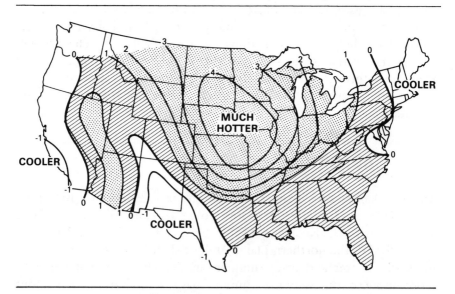

Table 9-1.   Monthly Averages, Havre, Montana.

Month	Temperature (Degrees F) 1930– 1939	1941– 1970[b]	Precipitation[a] (Inches) 1930– 1939	1941– 1970[b]	Snow (Inches) 1930– 1939	1941– 1970[b]
January	16.7	11.3	0.45	0.52	6.4	7.8
February	18.1	17.6	0.37	0.40	5.0	5.9
March	30.2	26.5	0.71	0.49	8.8	5.6
April	45.5	43.6	0.83	1.02	2.7	5.4
May	57.0	54.9	1.22	1.48	0.1	0.2
June	64.3	62.1	2.70	2.55		0.1
July	72.7	69.9	1.22	1.38		
August	68.6	68.0	1.18	1.05		
September	58.1	57.1	0.81	1.11	0.9	0.4
October	46.0	46.6	0.66	0.67	1.2	1.5
November	32.4	30.0	0.44	0.46	4.9	5.5
December	23.3	18.2	0.59	0.42	10.1	5.9
Annual	44.4	42.2	11.18	11.55	40.1	38.3

Source: Data from city location through 1930s.
[a] Includes melted snow.
[b] Current normals.

Lincoln's hottest day ever came in July 1936 as the mercury soared to 115°F. (Temperatures have exceeded 110°F five times in Lincoln since records were started in 1888, and all five times were in the 1930s.) Minneapolis–St. Paul's absolute maximum temperature occurred in July 1936, too, as did Bismarck's. Minneapolis–St. Paul tagged 108°F while Bismarck reached 114°F. In Valentine, the hottest day on record was in 1934, when a July day brought a reading of 110°F.

Concomitant with the broiling summers of the 1930s was a significant reduction in mean rainfall for the combined months of June, July, and August. At four of the five stations studied in the northern plains, totals for the three-month period ranged from 72 to 80 percent of the current average. In terms of rainfall, that meant deficits ranging from 1.5 inches at Bismarck to 3.5 inches at Minneapolis–St. Paul. Although July averaged somewhat drier at Havre during the

Table 9-2. Monthly Averages, Bismarck, North Dakota.

Month	Temperature (Degrees F)		Precipitation [a] (Inches)		Snow (Inches)	
	1930–1939	1941–1970[b]	1930–1939	1941–1970[b]	1930–1939	1941–1970[b]
January	9.8	8.2	0.34	0.51	5.2	7.3
February	13.9	13.5	0.52	0.44	7.6	6.3
March	28.0	25.1	0.65	0.73	6.0	7.9
April	44.1	43.0	1.10	1.44	3.0	4.0
May	57.1	54.4	1.83	2.17	0.7	1.3
June	66.5	63.8	2.97	3.58		
July	74.6	70.8	2.19	2.20		
August	70.5	69.2	1.04	1.96		
September	60.1	57.5	1.13	1.32	0.3	0.3
October	45.4	46.8	0.82	0.80	1.3	1.1
November	30.0	28.9	0.48	0.56	5.7	4.8
December	18.2	15.6	0.33	0.45	5.6	6.0
Annual	43.2	41.4	13.40	16.16	35.4	39.0

Source: Data from city location through 1930s.

[a] Includes melted snow.

[b] Current normals.

1930s, overall summertime precipitation there was about the same then as now.

Typically, the largest mean monthly deficits in summer rainfall during 1930–1939—as compared to now—occurred in either June or July (except at Bismarck, where it occurred in August). The maximum mean reduction for an individual month was at Minneapolis–St. Paul where July precipitation averaged just 52 percent of the current normal. And at Bismarck, August precipitation averaged only 53 percent of the modern mean.

Average annual precipitation during the 1930s, relative to now, diminished by as much as 21 percent at Lincoln, while only a minimal drop in yearly totals was apparent at Havre. At all five stations, however, from seven to nine of the ten years in the 1930s had less precipitation than the current normal. In Lincoln, only in 1931 was

Table 9-3.   Monthly Averages, Minneapolis-St. Paul, Minnesota.

| | Temperature | | Precipitation[a] | | Snow | |
| | (Degrees F) | | (Inches) | | (Inches) | |
Month	1930–1939	1941–1970[b]	1930–1939	1941–1970[b]	1930–1939	1941–1970[b]
January	14.8	12.9	0.97	0.73	9.4	8.2
February	18.1	17.2	0.88	0.84	7.8	7.9
March	30.3	29.0	1.43	1.68	6.8	10.4
April	45.3	45.8	1.88	2.04	2.1	2.4
May	60.1	57.8	3.69	3.37	0.3	0.2
June	70.0	67.8	3.48	3.94		
July	76.6	72.9	1.92	3.69		
August	72.9	71.1	2.80	3.05		
September	63.6	60.6	2.57	2.73		0.1
October	49.5	50.7	1.92	1.78	0.6	0.4
November	33.7	33.0	1.41	1.20	2.9	5.2
December	20.8	19.4	0.93	0.89	7.5	8.6
Annual	46.3	44.9	23.88	25.94	37.4	43.4

Source: Data from city location through 1937 with thermometer at 105 feet above ground.

[a] Includes melted snow.

[b] Current normals.

there enough rain and melted snow, 34.3 inches, to exceed the modern normal of 29.2 inches.

A number of city and state records for dryness were established during the dusty 1930s. In Montana, Helena's driest year ever was 1935 when only 6.28 inches of precipitation accumulated. State marks for absolute minimum annual precipitation were tallied in North Dakota in 1934—4.02 inches at Parshall in the northwest part of the state—and in South Dakota in 1936—2.89 inches at Ludlow in the northwest corner of the state.

City minimum marks for yearly precipitation were set in 1936 at both Fargo, North Dakota (8.87 inches), and Rapid City, South Dakota (7.51 inches). Williston, North Dakota's dustiest year on record was 1934 with just 6.13 inches of rain and melted snow. To the

Table 9-4.   Monthly Averages, Valentine, Nebraska.

Month	Temperature (Degrees F) 1930–1939	Temperature (Degrees F) 1941–1970[b]	Precipitation[a] (Inches) 1930–1939	Precipitation[a] (Inches) 1941–1970[b]	Snow (Inches) 1930–1939	Snow (Inches) 1941–1970[b]
January	21.6	20.4	0.50	0.31	5.5	4.1
February	25.1	24.5	0.45	0.53	6.0	5.7
March	34.2	31.2	1.29	0.76	10.2	6.8
April	46.9	45.8	2.28	1.77	3.6	3.1
May	58.6	56.9	2.92	2.80	0.6	0.7
June	69.7	66.9	2.22	3.60		
July	78.2	74.1	1.95	2.50		
August	74.1	72.5	1.94	2.38		
September	64.4	61.4	0.78	1.48		0.2
October	50.1	50.0	0.95	0.92	2.4	0.5
November	35.9	34.6	0.32	0.45	4.7	3.3
December	27.2	24.4	0.35	0.30	3.2	4.0
Annual	48.8	46.9	15.95	17.80	36.2	28.4

Source: Data from city location through 1930s with thermometers at 47 feet above ground.

[a] Includes melted snow.

[b] Current normals.

east, a state record for Minnesota was entered into the books in 1936 when the community of Angus measured 7.81 inches of annual rain and snowfall. Nebraska's record for lack of precipitation was set in 1931 at a location near Hull where only 6.30 inches fell. And city records were tallied in Omaha in 1934 (14.9 inches) and in North Platte in 1931 (10.01 inches).

Table 9–5. Monthly Averages, Lincoln, Nebraska.

Month	Temperature (Degrees F)		Precipitation[a] (Inches)		Snow (Inches)	
	1930–1939	1941–1970[b]	1930–1939	1941–1970[b]	1930–1939	1941–1970[b]
January	25.3	23.8	0.84	0.78	7.3	6.2
February	28.9	29.2	0.77	1.12	6.0	6.3
March	39.9	37.5	1.43	1.67	4.5	6.7
April	52.6	52.1	1.74	2.61	0.3	1.1
May	63.7	63.0	3.09	3.65		0.1
June	74.7	72.6	3.26	5.37		
July	82.4	78.1	2.68	3.60		
August	78.2	76.5	3.11	3.45		
September	69.6	66.6	2.67	3.32		
October	55.9	56.3	1.17	1.73		0.2
November	41.1	40.4	1.65	1.02	1.1	2.4
December	31.0	28.9	0.78	0.88	2.7	6.6
Annual	53.6	52.1	23.19	29.20	21.9	29.6

Source: Data from city location.

[a] Includes melted snow.

[b] Current normals.

Table 9−6.   Monthly Averages, Dodge City, Kansas.

Month	Temperature (Degrees F)		Precipitation[a] (Inches)		Snow (Inches)	
	1930−1939	1941−1970[b]	1930−1939	1941−1970[b]	1930−1939	1941−1970[b]
January	31.9	30.8	0.47	0.50	3.0	3.7
February	36.2	35.2	0.63	0.63	4.1	3.4
March	43.9	41.2	0.92	1.13	2.8	5.4
April	54.6	54.0	1.52	1.71	1.5	0.6
May	64.4	64.0	2.92	3.13		
June	75.6	73.7	2.12	3.34		
July	82.6	79.2	1.49	3.08		
August	80.6	78.1	1.79	2.64		
September	71.4	68.9	1.61	1.67		
October	58.2	57.9	1.17	1.65	0.1	0.2
November	43.8	42.8	0.64	0.59	2.6	2.3
December	35.7	33.4	0.43	0.51	1.6	3.4
Annual	56.6	54.9	15.71	20.58	15.7	19.0

Source: Data from city location through 1930s.

[a] Includes melted snow.

[b] Current normals.

THE SOUTHERN PLAINS AND MISSOURI

Across the southern plains (except for Texas) and in Missouri during the 1930s, the climatic trends evident on the northern plains were echoed. Winters were less cold than now, and summers appreciably hotter and drier. Overall, mean annual temperatures were about 2°F higher than modern averages.

Julys were palpably hotter across much of Kansas and Missouri, averaging 3.4°F above the current normal at Dodge City, Kansas (Table 9−6), 3.0°F above at Oklahoma City, Oklahoma (Table 9−7), and 4.3°F above at Columbia, Missouri (Table 9−8). (See Figure 9−1). Oklahoma City's highest temperature ever, 113°F, was in August, 1936.

Table 9—7. Monthly Averages, Oklahoma City, Oklahoma.

Month	Temperature (Degrees F)		Precipitation[a] (Inches)		Snow (Inches)	
	1930–1939	1941–1970[b]	1930–1939	1941–1970[b]	1930–1939	1941–1970[b]
January	38.8	36.8	1.73	1.11	1.6	3.0
February	43.0	41.3	1.25	1.32	0.5	2.3
March	51.1	48.2	1.93	2.05	0.8	2.2
April	61.0	60.4	2.08	3.47	0.4	
May	69.1	68.3	4.33	5.20		
June	78.9	76.8	4.86	4.22		
July	84.5	81.5	1.07	2.66		
August	83.6	81.1	3.26	2.56		
September	76.1	73.0	2.62	3.55		
October	63.9	62.4	2.48	2.57		
November	49.4	49.2	2.43	1.40	0.2	0.3
December	41.7	40.0	1.72	1.26	1.4	1.5
Annual	61.8	59.9	29.76	31.37	4.9	9.3

Source: Data from city location through 1930s.

[a] Includes melted snow.

[b] Current normals.

Large reductions in July precipitation accompanied the heat during the 1930–1939 period. Considered all together, the reductions are astounding. At Dodge City, July rainfall averaged only 48 percent of the modern mean, and at Columbia 53 percent. At Oklahoma City, just 40 percent of the modern average fell, with the mean precipitation for the month reduced to just over an inch. The current normal is over 2.5 inches.

While increased average precipitation during the other months of the year was able to make up a good deal of the July deficit in both Oklahoma City and Columbia, Dodge City, near the heart of the dust bowl, saw its annual precipitation during the 1930s dwindle to just 75 percent of the modern normal. Every single year of the 1930–1939 decade there brought less precipitation than the current average.

Table 9–8.   Monthly Averages, Columbia, Missouri.

Month	Temperature (Degrees F) 1930–1939	1941–1970[b]	Precipitation[a] (Inches) 1930–1939	1941–1970[b]	Snow (Inches) 1930–1939	1941–1970[b]
January	32.3	29.3	2.10	1.57	3.2	4.1
February	35.2	33.6	1.59	1.72	5.0	4.9
March	43.9	41.7	2.67	2.58	4.3	4.8
April	55.1	55.0	2.43	3.83	0.6	0.4
May	65.4	64.4	4.75	4.68		
June	75.1	73.0	3.80	4.59		
July	81.6	77.3	2.08	3.89		
August	78.6	76.0	3.94	3.19		
September	71.1	68.3	4.14	4.39		
October	58.5	58.0	2.31	3.38		
November	44.3	43.9	3.17	1.79	1.8	0.9
December	35.2	32.8	2.05	1.78	2.5	4.5
Annual	56.4	54.4	35.03	37.39	17.4	19.6

Source: Data from city location through 1930s, now airport at 150 feet higher elevation.
[a] Includes melted snow.
[b] Current normals.

TEXAS

Paradoxically, not too far south of Dodge City, the 1930s were relatively damp in Texas—at least below the panhandle. Data from Midland–Odessa, Texas (Table 9–9), Waco, Texas (Table 9–10), and Galveston, Texas (Table 9–11), tell us that much of west Texas—like New Mexico—was somewhat cooler and damper through the 1930s compared to now; while east Texas, on the other hand, was a bit milder with little change in precipitation. The relative mildness across eastern Texas was especially noticeable in winter. The January average was 1.5° to 2°F above modern normals.

Significantly hotter summer weather during the 1930s in the Lone Star State was apparently confined to the panhandle, closer to the

Table 9-9.   Monthly Averages, Midland-Odessa, Texas.

Month	Temperature (Degrees F)		Precipitation[a] (Inches)		Snow (Inches)	
	1930–1939	1941–1970[b]	1930–1939	1941–1970[b]	1930–1939	1941–1970[b]
January	44.5[c]	43.6	0.96[c]	0.59	0.1[c]	1.3
February	47.6[c]	47.8	0.97[c]	0.56	0.1[c]	0.7
March	54.5[c]	54.3	0.36[c]	0.59	0.4[c]	0.5
April	62.8[c]	64.3	0.98[c]	0.85		
May	71.1[c]	72.3	1.81[c]	2.16		
June	79.7	79.9	1.55	1.49		
July	80.9	82.3	2.09	1.82		
August	80.6	81.8	1.58	1.52		
September	74.9	75.4	2.32	1.54		
October	65.2	65.8	1.18	1.38		
November	51.4	53.3	0.69	0.49		0.2
December	44.6	45.9	0.85	0.52	0.3	0.7
Annual	63.2	63.9	15.34	13.51	0.9	3.4

[a] Includes melted snow.

[b] Current normals.

[c] 1931–1939.

real dust bowl. Amarillo was blistered by its highest summer average on record in 1934, when the combined mean for June, July, and August reached 81.7°F.

Elsewhere, summers averaged from slightly hotter (east Texas) to almost a degree cooler (Midland-Odessa) through the 1930s compared to now. Still, a number of stations noted their highest single reading ever between 1930 and 1939. Galveston hit 101°F in July 1932, Corpus Christi reached 105°F in July 1934, Fort Worth soared to 112°F in August 1936, Lubbock reached 109°F in June 1939, and Victoria topped that by one degree (110°F) in July of the same year.

Annual precipitation during the 1930s in west Texas, at least around Midland-Odessa, averaged about 14 percent greater than the current mean, which is identical to the departure observed at Albuquerque. And mirroring the trend that occurred in Arizona and New

Table 9-10.   Monthly Averages, Waco, Texas.

Month	Temperature (Degrees F) 1930–1939	1941–1970[b]	Precipitation[a] (Inches) 1930–1939	1941–1970[b]	Snow (Inches) 1930–1939	1941–1970[b]
January	48.1	47.0	3.10	1.87	0.4	0.8
February	51.9	50.9	2.61	2.38		0.4
March	58.8	57.2	2.82	2.36	0.1	0.1
April	67.3	67.3	2.83	4.02		
May	74.7	74.5	5.19	4.60		
June	82.7	81.9	2.82	2.73		
July	86.1	85.6	1.90	1.47		
August	85.9	85.7	1.19	1.81		
September	79.9	78.9	3.56	3.19		
October	69.5	69.1	2.26	2.55		
November	56.3	57.5	2.28	2.27	0.6	
December	49.8	49.8	2.90	2.01	1.4	0.1
Annual	67.6	67.1	33.46	31.26	2.5	1.4

Source: Data from city location through 1930s.

[a] Includes melted snow.

[b] Current normals.

Mexico, 1930 Septembers at Midland–Odessa were often significantly wetter than more modern ones. Farther north in west Texas, in the panhandle, mean yearly precipitation amounts through the dust bowl decade were probably noticeably less than modern means. This conclusion is based on Figure 11–1 in Chapter 11.

Precipitation over eastern Texas in the 1930s was little different from now, with means ranging from slightly more to slightly less than current averages. Januarys were certainly damper there during the 1930s than they are now, while (at least around Galveston) Junes were, on the whole, much drier. Galveston measured only .01 inch of rain in two different Junes —1930 and 1934.

In a word, most of the Great Plains and Missouri were significantly warmer and drier through the 1930s than in modern times. By contrast, much of Texas was a bit moister, and at least the southwestern regions were somewhat cooler.

Table 9-11. Monthly Averages, Galveston, Texas.

Month	Temperature		Precipitation[a]	
	1930–1939	1941–1970[b]	1930–1939	1941–1970[b]
January	55.9	53.9	4.54	3.02
February	57.7	56.2	2.97	2.67
March	61.7	61.0	2.53	2.60
April	68.4	69.2	2.68	2.63
May	75.7	75.9	2.62	3.16
June	81.8	81.3	1.44	4.05
July	83.3	83.2	4.59	4.41
August	83.7	83.3	4.06	4.40
September	80.8	80.0	4.55	5.60
October	73.8	73.1	2.79	2.83
November	62.7	63.5	3.54	3.16
December	57.0	57.1	4.26	3.67
Annual	70.2	69.8	40.57	42.20

Source: Data from city location.

[a] Includes melted snow.

[b] Current normals.

Maximum drying during the dust bowl decade centered on Kansas, Nebraska, and Iowa, while scorching hot Julys blistered South Dakota, Nebraska, Iowa, Kansas, and Missouri. At many stations in the midsection of the nation, average June or July (August at Bismarck) rainfall during the 1930s was roughly half what it is now! For our Midwestern farmers, the return of dust bowl-type weather is not a pleasant thing to contemplate.

10 THE GREAT LAKES, THE OHIO VALLEY, AND THE LOWER MISSISSIPPI VALLEY — THE 1930s

The comparative warmth and dryness of the 1930s lessened eastward from the heart of the Midwest, but significant droughty conditions still plagued the area around Detroit, Michigan — southern Michigan, northern Ohio, and northeastern Indiana.

THE GREAT LAKES AND THE OHIO VALLEY

Besides Detroit records (Table 10−2), records from Sault Sainte Marie, Michigan (Table 10−1), Rockford, Illinois (Table 10−3), and Cincinnati, Ohio (Table 10−4), were examined to get a feel for the weather of the 1930s across the Great Lakes region and through the Ohio Valley.

√ Average annual temperatures there were higher in the 1930s than now, with departures relatively more extreme over the western portions of the area. Januarys, in particular, were frequently much milder than currently. But summers, too, were hotter and drier than modern ones. Julys were notably hotter than now, with temperatures averaging anywhere from 1.5°F to about 3.0°F above current means. And in Michigan, Augusts were equally as much warmer and drier.

Record temperature highs were established at numerous Great Lake and Ohio Valley locations during the 1930s:

Location	Record Maximum	Date
Cairo, IL	106	August 1930
Chicago, IL	105	July 1934
Moline, IL	111	July 1936
Peoria, IL	113	July 1936
Rockford, IL	112	July 1936
Evansville, IN	107	July 1936
Fort Wayne, IN	106	July 1936
Indianapolis, IN	107	July 1934
South Bend, IN	109	July 1934
Terre Haute, IN	110	July 1936
Lexington, KY	108	July 1936
Alpena, MI	104	July 1936
Detroit, MI	105	July 1934
Grand Rapids, MI	108	July 1936
Cincinnati, OH	109	July 1934
Columbus, OH	106	July 1936
Sandusky, OH	105	July 1936
Toledo, OH	105	July 1936
Charleston, WV	108	July 1931
Huntington, WV	108	July 1930
Green Bay, WI	104	July 1936
La Crosse, WI	108	July 1936
Madison, WI	107	July 1936
Milwaukee, WI	105	July 1934

While most months of the year around the Great Lakes and Ohio Valley had warmer mean temperatures in the 1930s than now, Aprils and Octobers typically ran a bit cooler.

On an annual basis, precipitation over the region during 1930–1939 averaged somewhat less than currently. Detroit experienced the largest reduction, compared to modern means, with average yearly rain and snowfall during the 1930s amounting to about 88 percent of the current normal. That meant an average year in the 1930s had about four inches less precipitation than now. Eight out of ten years during the decade at Detroit were drier than a "normal" modern year. At Rockford, seven such years occurred during the 1930s.

On the whole, summers in the 1930s were significantly drier than more recent ones throughout the entire region. The combined aver-

Table 10-1.   Monthly Averages, Sault Sainte Marie, Michigan.

Month	Temperature (Degrees F) 1930–1939	Temperature (Degrees F) 1941–1970[b]	Precipitation[a] (Inches) 1930–1939	Precipitation[a] (Inches) 1941–1970[b]	Snow (Inches) 1930–1939	Snow (Inches) 1941–1970[b]
January	17.0	14.2	2.54	1.92	25.1	23.4
February	13.8	15.2	1.88	1.48	20.6	17.4
March	23.1	24.0	1.86	1.74	13.3	14.0
April	36.9	38.2	1.95	2.22	3.4	4.3
May	50.7	49.0	2.35	3.01		0.7
June	59.5	58.7	3.20	3.31		
July	65.7	63.8	2.18	2.60		
August	64.9	63.2	2.39	3.10		
September	56.9	55.3	3.93	3.85		0.2
October	45.2	46.2	3.26	2.85	2.5	2.3
November	32.9	32.8	3.36	3.26	15.6	14.8
December	21.9	20.1	2.29	2.36	21.3	25.7
Annual	40.7	40.0	31.19	31.70	101.8	102.8

Source: Data from city location through 1930s, now airport at 114 feet higher elevation.

[a] Includes melted snow.

[b] Current normals.

age rainfall for the months of June, July, and August ranged from 76 percent of the current normal at Detroit, to 93 percent of the current mean at Rockford.

State records for annual dryness were set in Indiana at Brookville in 1934 (18.67 inches), in Michigan at Croswell in 1936 (15.64 inches), in West Virginia at Upper Tract in 1930 (9.5 inches), and in Wisconsin at Plum Island in 1937 (9.5 inches).

Snowfall measurements from the 1930s suggest that winters averaged a bit snowier then than now around Detroit, less snowy in Rockford, and about the same near Sault Sainte Marie and Cincinnati.

Table 10-2.   Monthly Averages, Detroit, Michigan.

Month	Temperature (Degrees F)		Precipitation[a] (Inches)		Snow (Inches)	
	1930–1939	1941–1970[b]	1930–1939	1941–1970[b]	1930–1939	1941–1970[b]
January	27.8	24.6	2.08	1.91	7.9	8.1
February	27.4	26.6	1.97	1.75	7.5	7.2
March	35.4	35.3	1.99	2.47	7.3	5.5
April	46.1	47.7	3.02	3.22	1.2	1.1
May	59.8	58.1	3.21	3.31		
June	69.2	68.3	3.18	3.42		
July	73.7	72.3	2.15	3.10		
August	72.1	70.8	2.08	3.28		
September	65.0	63.6	2.68	2.16		
October	52.1	53.1	1.92	2.48	0.2	
November	39.5	40.1	1.83	2.32	3.2	2.5
December	29.2	28.5	1.69	2.27	8.3	7.0
Annual	49.8	49.1	27.80	31.69	35.6	31.4

Sources: Snow data from city locations; other data from suburban locations.

[a] Includes melted snow.

[b] Current normals.

Table 10-3.   Monthly Averages, Rockford, Illinois.

Month	Temperature (Degrees F) 1930–1939	Temperature (Degrees F) 1941–1970[b]	Precipitation[a] (Inches) 1930–1939	Precipitation[a] (Inches) 1941–1970[b]	Snow (Inches) 1930–1939	Snow (Inches) 1941–1970[b]
January	24.7	20.2	2.00	1.79	6.5	9.1
February	26.3	24.0	1.21	1.29	5.2	6.3
March	36.1	34.1	2.12	2.65	7.1	7.3
April	48.0	48.2	2.25	3.85	1.5	0.9
May	61.5	58.8	3.65	3.86	0.1	
June	71.0	68.8	4.19	4.42		
July	76.5	72.8	3.36	4.27		
August	73.6	71.5	3.89	3.66		
September	65.9	63.3	4.10	4.00		
October	52.6	52.7	2.79	2.85	0.5	0.1
November	39.1	37.6	2.49	2.37	0.9	3.1
December	27.3	24.9	1.42	1.71	6.0	7.6
Annual	50.2	48.1	33.47	36.72	27.8	34.4

Source: Data from city location through 1930s.

[a] Includes melted snow.

[b] Current normals.

Table 10–4.   Monthly Averages, Cincinnati, Ohio.

Month	Temperature (Degrees F)		Precipitation[a] (Inches)		Snow (Inches)	
	1930–1939	1941–1970[b]	1930–1939	1941–1970[b]	1930–1939	1941–1970[b]
January	35.1	32.1	3.80	3.40	4.0	5.6
February	35.5	34.4	2.05	2.95	4.2	4.0
March	42.5	42.9	4.10	4.13	3.3	3.9
April	52.9	55.1	3.11	3.85	0.3	0.3
May	64.3	64.4	3.40	3.96		
June	73.7	73.1	3.39	3.92		
July	78.1	76.2	3.37	3.96		
August	75.8	75.1	3.25	3.02		
September	69.9	68.4	3.47	2.69		
October	56.6	57.8	1.96	2.20		
November	44.7	44.6	2.23	3.08	2.2	1.9
December	35.2	34.4	2.41	2.87	4.1	4.3
Annual	55.4	54.9	36.54	40.03	18.1	20.0

Source: Data from Abbe Observatory.
[a] Includes melted snow.
[b] Current normals.

THE LOWER MISSISSIPPI VALLEY

The 1930s in the lower Mississippi Valley averaged about 1°F milder than currently, but displayed little variation from current means in average yearly precipitation.

Annual precipitation totals at Memphis, Tennessee (Table 10–5), during the 1930s were down slightly (by 5 percent)—compared to the current average—but were up a bit (by 2 percent) at Baton Rouge, Louisiana (Table 10–6).

At both Memphis and Baton Rouge, Januarys in the 1930s averaged markedly milder and wetter than now, while Aprils were typically cooler and drier.

Table 10-5.   Monthly Averages, Memphis, Tennessee.

Month	Temperature (Degrees F)		Precipitation[a] (Inches)		Snow (Inches)	
	1930–1939	1941–1970[b]	1930–1939	1941–1970[b]	1930–1939	1941–1970[b]
January	43.6	40.5	7.37	4.93	1.0	2.4
February	45.9	43.8	4.32	4.73	1.0	1.2
March	52.8	51.0	4.83	5.10		1.0
April	61.9	62.5	3.23	5.42		
May	71.1	70.9	3.86	4.39		
June	79.1	78.6	2.72	3.46		
July	82.4	81.6	3.61	3.53		
August	81.5	80.4	1.53	3.33		
September	76.3	73.6	2.87	3.01		
October	64.6	63.0	2.64	2.58		
November	51.8	50.9	4.26	3.92	0.6	
December	44.2	42.7	5.30	4.70	1.0	1.0
Annual	62.9	61.6	46.54	49.10	3.6	5.6

Source: Data from city location through 1930s with extreme thermometers at 78 feet.
[a] Includes melted snow.
[b] Current normals.

Average summertime precipitation over the region during the dust bowl decade—compared to now—ranged from almost 25 percent less at Memphis, to about 7 percent greater at Baton Rouge. Average August rainfall at Memphis during the 1930–1939 period was only 46 percent of the current mean, while closer to the Gulf of Mexico, Baton Rouge was appreciably wetter during the same month.

Near the Mississippi River, Jackson, Mississippi, reached its all-time high temperature mark—107°F—in July 1936; in the same year, Yazoo City was in the process of setting a state mark for annual dryness with just 25.97 inches of precipitation. Obviously, while the lower end of the Mississippi Valley was spared the more persistent heat and drought of the 1930s, it was not immune to occasional incursions.

In summary, while areas just to the east of the midsection of the nation shared to some extent the Great Plains' heat of the 1930s,

Table 10-6.   Monthly Averages, Baton Rouge, Louisiana.

Month	Temperature (Degrees F)		Precipitation[a] (Inches)	
	1930–1939	1941–1970[b]	1930–1939	1941–1970[b]
January	55.2	51.0	6.24	4.40
February	57.4	53.9	4.32	4.76
March	61.0	59.7	3.58	5.14
April	67.2	68.4	4.36	5.10
May	74.3	74.8	5.34	4.39
June	80.5	80.3	3.39	3.77
July	82.0	82.0	6.54	6.51
August	81.6	81.6	6.14	4.67
September	78.5	77.5	3.82	3.79
October	69.7	68.5	3.12	2.65
November	59.1	58.6	3.92	3.84
December	54.5	52.9	4.55	5.03
Annual	68.4	67.4	55.32	54.05

Source: Data from city location through 1930s.

[a] Includes melted snow.

[b] Current normals.

they did not necessarily partake of the extreme drought. Average annual precipitation over most of the region from the Great Lakes southward in the 1930s was less than 10 percent lower than current normals, except near Detroit where there was some chronic dryness.

Summers over most of the area—excluding southern Louisiana—were generally hotter and drier in the 1930s than currently. But the greatest amount of mean warming, compared to now, was reserved for the month of January. January was also a wetter month than now over the region—markedly so in the lower Mississippi Valley. Aprils, on the other hand, were normally cooler and drier than more recent ones.

Finally, as we shall see in the following chapter, the manifestations of the extreme heat and drought of the 1930s disappeared entirely along the country's eastern seaboard.

11 THE EASTERN UNITED STATES— THE 1930s

If you had lived in one of the major metropolitan areas of the East during the 1930s, and had, for some reason, been deprived of access to radio broadcasts, newspapers, and movie newsreels, you probably would not have been aware that the 1930s were climatically much different from any other decade.

From Boston to Washington, D.C., the 1930s—compared to today's means—ranged from slightly cooler to unchanged in terms of mean annual temperature. And yearly precipitation totals, on the average, were actually a bit greater than now.

The dust bowl of the 1930s was far removed from the eastern United States.

THE NORTHEAST

Data from Burlington, Vermont (Table 11−1), Rochester, New York (Table 11−2), Milton (near Boston), Massachusetts (Table 11−3), and New York, New York (Table 11−4), tell us that the 1930s were milder and drier over western portions of the northeastern United States, and cooler and slightly damper to the east, than they are now.

Rochester displayed the largest mean warming and the greatest reduction in average annual precipitation in the Northeast during the

109

Table 11-1.   Monthly Averages, Burlington, Vermont.

Month	Temperature (Degrees F)		Precipitation[a] (Inches)		Snow (Inches)	
	1930–1939	1941–1970[b]	1930–1939	1941–1970[b]	1930–1939	1941–1970[b]
January	20.0	16.8	2.22	1.74	15.8	17.0
February	18.8	18.6	1.54	1.68	12.7	16.9
March	28.5	29.1	2.48	1.93	12.2	11.7
April	42.0	43.0	2.78	2.62	5.0	2.4
May	55.4	54.8	2.73	3.01		0.2
June	65.1	65.2	3.47	3.46		
July	69.5	69.8	3.92	3.54		
August	67.8	67.4	2.89	3.72		
September	59.8	59.3	3.39	3.05		
October	48.3	48.8	2.98	2.74	0.1	0.2
November	35.9	37.0	2.08	2.86	4.0	6.2
December	22.9	22.6	2.10	2.19	12.8	18.0
Annual	44.5	44.4	32.58	32.54	62.6	72.6

Source: Data from city location through 1930s.

[a] Includes melted snow.

[b] Current normals.

1930s compared to now; while on the other hand, Milton (Boston) had the most significant cooling, and New York City the biggest increase in mean yearly rain and snowfall.

Annual precipitation at Rochester averaged about 7 percent less in the 1930s than in more modern times, while yearly totals in New York City were up roughly 11 percent. Even bigger differences in mean annual snowfall amounts were apparent in the 1930s, compared to now, with significant reductions evident at all the northeastern stations studied. Around Boston annual snowfall totals during the 1930–1939 decade were only 73 percent of the modern normal, while lesser reductions occurred at Burlington and New York City. There the 1930s averages were 86 percent of the current normals.

One of the reasons for the diminished snowfall of the 1930s— particularly at Rochester and in the Boston area—was that fact that,

Table 11-2. Monthly Averages, Rochester, New York.

Month	Temperature (Degrees F)		Precipitation[a] (Inches)		Snow (Inches)	
	1930–1939	1941–1970[b]	1930–1939	1941–1970[b]	1930–1939	1941–1970[b]
January	28.3	24.0	2.66	2.25	14.9	21.6
February	25.9	24.8	2.07	2.42	13.9	21.7
March	33.6	33.0	3.41	2.57	15.7	14.1
April	44.5	46.1	2.20	2.74	2.6	2.2
May	58.2	56.5	2.24	2.80		0.1
June	68.0	66.9	2.57	2.54		
July	73.0	71.2	2.24	2.89		
August	71.1	69.3	2.71	2.97		
September	63.9	62.3	2.51	2.35		
October	52.1	52.3	2.00	2.62	0.4	0.4
November	40.7	40.5	2.09	2.83	5.6	6.9
December	30.3	28.3	2.33	2.35	12.7	16.2
Annual	49.1	47.9	29.03	31.33	65.8	83.2

Source: Data from city location through 1930s with thermometers at 86 feet.
[a] Includes melted snow.
[b] Current normals.

at all locations, Januarys were noticeably milder then, by an average of 2 to 4°F. Aprils and Octobers, on the other hand, were typically cooler, similar to the trend manifest around the Great Lakes.

While prolonged drought was not a problem in the Northeast during the dust bowl decade, the year 1930 did produce some parched conditions in New England, where minimum annual precipitation in both Maine and New Hampshire set state records. In Maine it happened "Down East" at Machias (23.06 inches), and in New Hampshire at Bethlehem (22.31 inches). It was also dusty in eastern Pennsylvania that same year. Wilkes-Barre/Scranton (26.12 inches) and Williamsport (27.68 inches) experienced their driest twelve months in history.

If I may digress for a moment, an interesting discovery of my study of the northeastern weather records was a remarkable increase in seasonal snowfall that has taken place around Rochester (and

Table 11–3.   Monthly Averages, Milton (Boston), Massachusetts.

Month	Temperature (Degrees F)		Precipitation[a] (Inches)		Snow (Inches)	
	1930–1939	1941–1970[b]	1930–1939	1941–1970[b]	1930–1939	1941–1970[b]
January	27.7	25.8	4.64	4.12	12.7	17.6
February	25.6	27.0	3.46	3.97	14.3	16.8
March	33.2	34.6	4.42	4.51	8.8	15.0
April	43.3	45.5	4.36	3.64	1.7	2.2
May	55.7	55.8	3.03	3.62		
June	63.8	64.9	4.46	3.15		
July	68.9	70.4	3.31	2.95		
August	68.3	68.6	4.22	3.83		
September	61.0	61.7	4.79	3.65		
October	50.4	52.6	4.24	3.62	0.4	0.2
November	39.7	41.7	3.40	5.06	3.5	2.3
December	28.8	29.4	3.57	4.70	7.2	12.4
Annual	47.2	48.2	47.90	46.82	48.6	66.5

Source: Data from Blue Hill Observatory.

[a] Includes melted snow.

[b] Current normals.

seemingly at other eastern Great Lakes locations, too) since the winter of 1955–56. From 1920 through 1955 the average annual snowfall at Rochester was 67.6 inches. From 1956 through 1978 the average is up by over 30 inches with the twenty-three-year mean at 101.4 inches! It is unlikely that the difference can be attributed to a change in weather station location, because the only move that could have significantly affected snowfall amounts at Rochester took place in 1940.

The eastern Great Lakes snowfalls perhaps reached a pinnacle in the winter of 1976–77 with the awesome snows around Buffalo, New York. Snowy winters will probably continue to be the norm around the eastern Great Lakes for awhile,[1] and we may have to await the advent of anthropogenic climatic change for the snows to diminish. But that is only speculation.

Table 11-4.   Monthly Averages, New York, New York.

Month	Temperature (Degrees F) 1930-1939	Temperature (Degrees F) 1941-1970[b]	Precipitation[a] (Inches) 1930-1939	Precipitation[a] (Inches) 1941-1970[b]	Snow (Inches) 1930-1939	Snow (Inches) 1941-1970[b]
January	34.7	32.2	3.98	2.71	6.4	6.8
February	32.5	33.4	3.12	2.92	7.6	7.5
March	40.2	41.1	4.13	3.73	3.0	5.8
April	49.8	52.1	3.46	3.30	0.6	0.7
May	62.9	62.3	3.23	3.47		
June	71.6	71.6	4.13	2.96		
July	76.7	76.6	3.89	3.68		
August	75.5	74.9	4.39	4.01		
September	68.7	68.4	5.15	3.27		
October	57.4	58.7	3.22	2.85		
November	46.2	47.4	3.05	3.76	2.2	0.4
December	35.9	35.5	2.86	3.53	4.6	7.1
Annual	54.3	54.5	44.61	40.19	24.4	28.3

Source: Data from Central Park.

[a] Includes melted snow.

[b] Current normals.

In the meantime we are limited to the knowledge that the snow climate of Rochester, and possibly of all western New York, did undergo a significant alteration in the mid-1950s, probably because of changing wind flows and storm tracks. Of course, no one is yet able to say for certain why the pattern of winds and storms might have changed.

Table 11–5. Monthly Averages, Washington, D.C.

Month	Temperature (Degrees F)		Precipitation[a] (Inches)		Snow (Inches)	
	1930– 1939	1941– 1970[b]	1930– 1939	1941– 1970[b]	1930– 1939	1941– 1970[b]
January	38.2	35.6	3.95	2.62	5.1	4.9
February	37.2	37.3	2.89	2.45	6.0	4.7
March	44.9	45.1	3.41	3.33	2.8	3.5
April	54.3	56.4	3.33	2.86		
May	65.7	66.2	3.86	3.68		
June	74.3	74.6	3.27	3.48		
July	78.4	78.7	3.56	4.12		
August	77.0	77.1	4.36	4.67		
September	70.5	70.6	5.08	3.08		
October	57.8	59.8	3.08	2.66		
November	47.8	48.0	2.54	2.90	0.9	0.7
December	38.3	37.4	2.62	3.04	3.8	4.0
Annual	57.0	57.3	41.95	38.89	18.6	17.8

[a] Includes melted snow.

[b] Current normals.

THE MIDDLE ATLANTIC STATES

The Middle Atlantic states in the 1930s were little different from now in terms of temperature. In terms of precipitation the area was a bit wetter over eastern sections, but neither significantly wetter nor drier closer to the Appalachians. Average annual precipitation at Washington, D.C. (Table 11–5), during the 1930s was 108 percent of the current mean, while at Greensboro, North Carolina (Table 11–6), mean yearly precipitation was about the same as it is now.

As was common over much of the eastern half of the country, Januarys averaged much milder during the 1930s than now at both Washington and Greensboro; Aprils and Octobers were often cooler — an echo of the pattern in the Northeast and throughout the Great Lakes.

Table 11-6.   Monthly Averages, Greensboro, North Carolina.

Month	Temperature (Degrees F) 1930-1939	Temperature (Degrees F) 1941-1970[b]	Precipitation[a] (Inches) 1930-1939	Precipitation[a] (Inches) 1941-1970[b]	Snow (Inches) 1930-1939	Snow (Inches) 1941-1970[b]
January	40.8	38.7	3.71	3.22	2.0	3.7
February	41.0	40.6	3.03	3.37	1.4	2.2
March	47.6	47.8	3.72	3.72	0.8	2.0
April	55.9	58.6	3.44	3.15		
May	66.6	67.1	3.65	3.04		
June	74.7	74.4	3.34	3.91		
July	77.7	77.2	4.90	4.39		
August	76.2	76.0	4.85	4.30		
September	71.3	69.7	2.94	3.55		
October	58.4	59.2	2.34	2.94		
November	47.7	48.3	2.57	2.62		0.2
December	39.6	39.6	3.49	3.15	2.9	0.9
Annual	58.1	58.1	41.98	41.36	7.1	9.0

[a] Includes melted snow.
[b] Current normals.

While drought did not plague the Middle Atlantic states during the 1930s, there were some dry periods early in the decade. State records for annual dryness were set in Maryland at Picardy (17.76 inches) in 1930 and in North Carolina at Mount Airy (22.69 inches) in the same year. City marks for a relative lack of yearly precipitation were set at several locations in 1930: Baltimore, Maryland (21.55 inches), Washington, D.C. (21.66 inches), Lynchburg, Virginia (19.83 inches), and Roanoke, Virginia (23.17 inches). Raleigh, North Carolina's, driest year ever came in 1933: 29.93 inches. On the other side of the Appalachians, residents of Knoxville, Tennessee, experienced the driest year in history there in 1930 when there were just 33.67 inches of rain and melted snow.

Still, overall precipitation during the 1930s was certainly adequate. And even though average annual totals along the coastal plain of the Middle Atlantic states up into southern New England were actually greater in the 1930s than now, a study completed in 1975 concluded that nontropical storminess along the Mid-Atlantic coast and just offshore was somewhat less during 1930–1939 than recently. About 5 percent fewer storms—or about 2 per year—occurred in the 1930s than in the 1960s. Even more important, the study found, the type of nontropical storm that most often generates the highest waves and storm surges along the coast occurred about 15 percent less frequently in the 1930s than in the 1960s.[2] Remember, though, that the mid-Atlantic region and New England were rocked by several severe *tropical* storms during the dust bowl decade (see Chapter 5).

THE SOUTHEAST

The southeastern United States throughout the 1930s averaged warmer than now and, at least over the Florida peninsula, slightly wetter. Historical records from Rome, Georgia (Table 11–7), and Apalachicola, Florida (Table 11–8), indicate that mean yearly precipitation through the 1930s was little different from currently. However, the record from Lakeland, Florida (Table 11–9), tells us that the 1930s there were somewhat rainier than more modern times, and that most of that enhanced rainfall came in the period February through June.

Winters over the Southeast in the 1930s were markedly warmer than current ones, with January acting as the archetype, as it had the biggest warming of winter months. Again, the relatively warm Januarys during the 1930s seemed to be a common phenomenon over much of the eastern half of the United States.

Another common phenomenon over the East and during January was higher mean precipitation than today, and this appeared in the records at both Rome and Apalachicola. Lack of precipitation did not present a chronic problem to the Southeast in the 1930–1939 period, but, as was the case in the mid-Atlantic states, there were some notably dry years early in the decade.

Savannah, Georgia, and Jacksonville, Florida, were hit with extreme dryness in 1931. Both cities established records for least annual precipitation ever; Savannah with 22 inches and Jacksonville with 34.38 inches. In 1933, Augusta, Georgia (28.05 inches), and

Table 11–7.   Monthly Averages, Rome, Georgia.

Month	Temperature (Degrees F)		Precipitation[a] (Inches)		Snow (Inches)	
	1930–1939	1941–1970[b]	1930–1939	1941–1970[b]	1930–1939	1941–1970[b]
January	47.1	41.1	6.02	4.99	N/A	0.8
February	47.2	44.0	4.94	5.19	N/A	0.6
March	53.3	50.5	5.39	6.17	N/A	0.4
April	61.2	60.7	5.07	4.77		
May	70.6	68.6	3.83	3.93		
June	77.8	75.8	4.06	3.74		
July	80.9	78.7	5.22	4.74		
August	79.7	78.1	4.15	3.49		
September	75.4	72.2	2.88	3.92		
October	63.1	61.0	3.43	2.90		
November	51.9	49.5	2.89	3.76		
December	44.8	42.0	5.66	4.99		0.4
Annual	62.7	60.2	53.54	52.59		2.2

[a] Includes melted snow.   N/A   Not available.

[b] Current normals.

Columbia, South Carolina (27.11 inches), joined Savannah and Jacksonville in the record book. Along the eastern Gulf Coast, Mobile, Alabama, had its driest year ever in 1938, with 37.15 inches of rain.

In summary, the eastern United States was relatively immune to the effects of the dust bowl decade. There were some dry years in the region early in the decade, but drought was never a persistent problem. In fact, average annual precipitation totals were significantly higher than they are now along much of the coastal plain from southern New England to eastern North Carolina. In the Northeast, the trend for snowfall was in the opposite direction. Seasonal totals averaged only about 80 percent of the current means.

Much of the Southeast was a bit warmer, on the average, during the 1930s, while the rest of the East had average annual temperatures very close to recent ones. Near the coast, the 1930s climate actually ran a bit cooler than now. But, the month of January was universally and significantly milder throughout the eastern United States.

Table 11-8.   Monthly Averages, Apalachicola, Florida.

Month	Temperature (Degrees F)		Precipitation[a] (Inches)	
	1930–1939	1941–1970[b]	1930–1939	1941–1970[b]
January	56.5	53.7	4.16	3.07
February	57.1	55.8	3.86	3.78
March	60.7	60.7	3.80	4.70
April	67.0	68.3	4.09	3.61
May	74.9	74.9	3.04	2.78
June	79.7	80.0	5.65	5.30
July	81.9	81.4	6.77	8.02
August	81.1	81.5	8.33	8.07
September	79.2	78.6	7.42	9.00
October	71.2	70.8	2.11	2.88
November	61.0	61.1	2.99	2.68
December	56.1	55.2	2.94	3.32
Annual	68.9	68.5	55.16	57.21

Source: Data from city location.

[a] Includes melted snow.

[b] Current normals.

If nothing else important has come to light in the last five chapters, it at least should have become clear that the effects of climatic warming are far from uniform. Many areas grow warmer, some cooler, some drier, and some wetter. We are not dealing with a simple pattern of straightforward changes.

By studying the complex differences (as compared to current climatic normals) that prevailed in the United States during the 1930s, we may at least get a general feeling for the initial effects of the greenhouse threat.

Finally, it should be pointed out that even similar climatic changes can have dissimilar effects on different segments of our society. Relatively little snow and cold in the Northeast in Januarys would mean homeowners could bank money saved by lowered heating costs, while ski areas teeter on the edge of bankruptcy. In other areas of

Table 11-9.   Monthly Averages, Lakeland, Florida.

Month	Temperature (Inches)		Precipitation[a] (Inches)	
	1930-1939	1941-1970[b]	1930-1939	1941-1970[b]
January	64.3	60.8	1.79	2.32
February	65.3	62.1	3.07	2.52
March	67.4	66.3	4.31	4.02
April	72.2	72.0	3.97	2.57
May	77.6	77.0	4.64	3.44
June	80.1	80.5	8.95	6.70
July	82.4	81.6	6.97	8.09
August	82.3	81.9	7.08	7.18
September	80.7	80.2	6.91	6.06
October	75.2	74.3	2.24	2.84
November	66.9	66.8	1.39	1.60
December	63.0	62.0	1.62	2.09
Annual	73.1	72.1	52.94	49.43

Source: Data from city location.

[a] Includes melted snow.

[b] Current normals.

the country, hot, sunny summers might be a boon to tourist and recreational industries, but could spell disaster for agriculture. And in the end, it is undoubtedly agriculture that bears the brunt of the suffering.

A REVIEW: THE DUST BOWL DECADE

The most severe desiccation of the 1930s plagued the center of the nation, particularly the area from eastern Colorado through much of Nebraska and Kansas into Iowa. There, mean annual precipitation amounts during 1930-1939 averaged 80 percent or less of current normals. Figure 11-1 shows the pattern of average yearly precipitation across the United States in the 1930s as a function (percentage) of current normals.

Figure 11–1. The pattern of average yearly precipitation in the 1930s as a function (percentage) of current normals. The most severe desiccation of the 1930s occurred in the center of the nation, particularly the area from eastern Colorado through much of Nebraska and Kansas into Iowa.

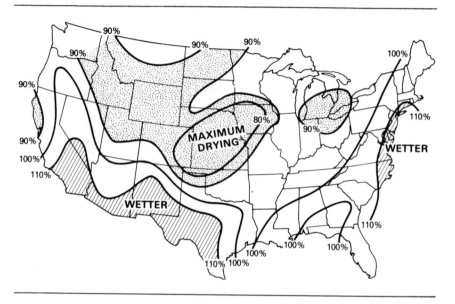

Even more striking—and potentially more devastating—than the pattern of annual precipitation changes was the scope of July drought across the Great Plains. In most of Kansas and Oklahoma—the heart of America's wheat-growing land—July rainfall during the 1930s averaged less than one-half the modern normal! Wheat yields, as was pointed out in Chapter 4, are significantly influenced by, among other factors, July precipitation.

Wheat yields, as well as corn crops, are also affected by July temperatures. And, as is readily apparent from Figure 9–1 in Chapter 9, July temperatures in many regions soared during the 1930s, blistering both wheat and corn belt states. South Dakota, Nebraska, Kansas, Missouri, and Iowa received the most severe heat, baking under the hottest summers in memory.

On an annual basis, the 1930s warming across the nation extended from the northern plains southeastward into Florida, with a pocket of warming in the Southwest through much of Arizona and Nevada.

Figure 11−2 shows the extent of the warming (and cooling) relative to modern normals.

I have discussed, and as is obvious from Figures 9−1, 11−1, and 11−2, dust bowl-type weather was not generic to the entire nation during the 1930s. The eastern seaboard averaged somewhat cooler and wetter, although Januarys were often significantly milder—as was the case throughout the eastern half of the country. In much of the Southwest, from New Mexico through Arizona into southern California, a typical year in the 1930s had precipitation significantly greater than current normals. But along the northern California coast, a typical 1930s year was probably a bit drier than now.

Most of the West Coast, as well as the East Coast, experienced a relative cooling during the dust bowl decade, as did New Mexico and southwestern Texas. Throughout much of the interior West, average

Figure 11−2.   The extent of the warming (and cooling) of the 1930s relative to modern normals. The most significant warming extended from the northern plains southeastward into Florida, with another area in the Southwest through much of Arizona and Nevada.

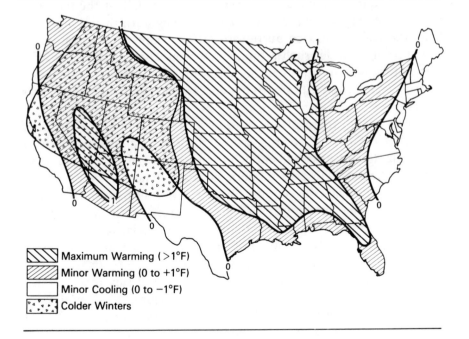

Maximum Warming (>1°F)
Minor Warming (0 to +1°F)
Minor Cooling (0 to −1°F)
Colder Winters

annual temperatures during 1930–1939 were pretty much unchanged from modern means, although winters averaged noticeably more chilly.

The patterns drawn in Figures 9–1, 11–1, and 11–2 have good coherence. That is, the "plus" and "minus" areas form a "logical" pattern, and do not end up as a bunch of unconnected cells. One can assume from this observation that some sort of large scale process must have been controlling the climatic changes of the 1930s. But do the patterns make any kind of meteorological sense? Beyond knowing that the westerlies shifted a bit northward during the 1930s, what kind of *specific* changes in the wind patterns over the United States can we envision that could have fostered drought on the plains, and heavier rains in southwest Texas, blistering hot July winds in Omaha, and cooling summer breezes in Boston, frigid winters in Salt Lake City, and warmer, wetter Januarys in New York City?

A winter time wind pattern that probably occurred more frequently in the 1930s than now is one that sweeps cold air out of British Columbia and Alberta directly into the intermountain West: a strong northerly flow. Meteorologists know that when this pattern sets up, the balancing reaction over the eastern half of the nation is often an intense southwesterly flow, whipping warmth and moisture from the Gulf of Mexico northward into the Great Lakes states and New England. Under such a broad scale wind configuration the West suffers through a colder-than-normal winter, or January, while much of the eastern United States experiences a comparatively balmy, wet January.

Across the northern plains, even in winters during the 1930s that had relatively normal steering winds, there was an apparent lack of Arctic air significantly cold enough to bring average readings, particularly in January, close to modern normals. (The winter of 1935–1936 was an obvious exception.) With the overall pattern of westerlies shifted somewhat northward in the 1930s, Arctic air was probably more often confined to the interior of Canada, at least in those winters when the icy air was not plunging into the western United States.

The pattern of average annual temperatures and July temperatures of the 1930s relative to current normals (Figures 11–1 and 9–1, respectively) are basically similar to one another, except for the extreme July anomalies that prevailed in the central part of the country. And as has been pointed out, a majority of the drying in the Midwest during the 1930s took place in July. Thus, by postulating a July wind flow configuration that may have prevailed more often during 1930–1939 than now, we can perhaps get a clue to the type of wind circulations that were responsible for a large part of the relative climatic change that took place then.

One particular circulation pattern that was probably dominant relatively more frequently during Julys of the 1930s is one that features a large area of high pressure aloft centered over the central or northern Great Plains. Near the center of high pressure aloft air gently descends, thus warming and drying under the influence of its own compression. At the surface under the high pressure area (or "high") hot, dry, sunny weather proliferates. If such a weather pattern becomes persistent over the Great Plains, drought and heat wave conditions develop. At the same time, the wind circulation around the high produces different kinds of weather in other areas of the country.

The winds around a high blow in a clockwise direction in the Northern Hemisphere. Therefore, in the example cited, northerly or northwesterly winds would prevail over the Northeast and Middle Atlantic states. Thus, relatively cooler air masses from Canada would be drawn into areas near the eastern seaboard. And a greater frequency of cold or cool frontal passages might generate increased precipitation—as was evidenced in the 1930s July rainfall averages at several stations in the Northeast—because of increased shower and thunderstorm activity along the fronts.

To continue with our example, winds along the southern periphery of a high over the Great Plains would blow from the east, forcing moisture from the Gulf of Mexico through southern Texas, then into the southwestern United States where the winds turn more southeasterly. The terrain from the Gulf of Mexico across western Texas into New Mexico rises steadily. Moisture-laden air masses following such a course are subject to lifting, which cools the air, condensing the moisture into clouds and rain.

So it becomes obvious that the same persistent upper air circulation that can produce drought in the Midwest can bring increased amounts of rainfall to much of southern Texas and the southwestern United States. Of course, where relatively greater concentrations of clouds and showers prevail, the amount of sunshine reaching the surface of the earth is diminished. It follows that surface air temperatures would be lowered from their normal levels. See Figures 9-1 and 11-2.

Once an air mass moving westward from the Gulf of Mexico crosses the continental divide, it descends, warming and drying. That fact could account for the relative July warming reported from Phoenix, Arizona, and Winnemuca, Nevada, during the 1930s. The two locations are downslope from nearby higher terrain.

Finally, as the winds along the western side of our model high turn southerly in the western United States, cooler Pacific Ocean air would be encouraged to flow into coastal areas, especially in southern California, where compared to more modern times there was significant July cooling observed in the 1930s.

In summary, it appears as though the pattern of relative climatic changes that occurred during the dust bowl decade do make meteorological sense. The changes were not universally of the same magnitude, nor even in the same direction. But they were the result of an altered atmospheric circulation pattern that went hand-in-hand with the hemispheric warming of the period. The alteration was not triggered by a greenhouse effect, but that is not important.

What we really wanted to look at were the manifestations of climatic warming, which, even for CO_2-induced warming, are likely to be initially similar to the natural warming of the 1930s. The lesson is there.

A HEMISPHERIC PERSPECTIVE

Shortly after I completed my analysis of U.S. climatic data of the 1930s as compared with current "normals," the results of some somewhat similar work were published in a British scientific journal. A group of researchers from the Climatic Research Unit, University of East Anglia, Norwich, United Kingdom, had examined differences in Northern Hemispheric temperature and precipitation determined by comparing the five warmest years in the period 1925-1974 with the five coldest years in the same period.[3] It was the contention of the researchers that such a comparison would provide a good climatic analog for a future high-CO_2 world.

The change in mean annual hemispheric temperature calculated by the British experiment amounts to about 1°F. That is, the five warmest years during 1925–1974 averaged roughly 1°F warmer than the five coldest. The 1°F figure is essentially how much warmer the 1930s were than now. Thus, it seems likely the patterns obtained in the British work should roughly parallel the ones I came up with for the United States. Beyond that point of interest, it is obvious that any research carried out on a hemispheric scale gives us a much broader picture, geographically speaking, of what the greenhouse effect might be like.

While it seems plausible there should be some general similarity between the results of the British effort and my analysis, a comparison of the two should not yield identical patterns in the United States. For one thing, the hemispheric analysis is based on comparing extreme years with extreme years (warm with cold). My analysis compares an anomalously warm decade with thirty-year "normals." And even within an unusually warm decade not all years are anywhere close to "extreme." So the British work probably overstates the warming (or cooling) and drying (or increased rainfall) according to my more conservative approach. Still, the comparison of patterns in the United States is interesting.

The British results indicate warming for the entire United States except for slight cooling along the California coast. The most important warming—up to about 3.5°F on a mean annual basis—extends across the northern Great Plains, then just south of the Great Lakes to the central Appalachians. The California cooling duplicates my results, but no eastern seaboard or Texas cooling is indicated. And the magnitude of the warming in the center of the nation is certainly greater than what I calculated, which was 1 to 2°F. Still, the patterns are broadly similar.

The mean annual precipitation patterns are also broadly similar. The British work indicates drying across much of the United States. But precipitation increases are delineated along the Pacific Coast and through the Pacific Northwest, in the lower and mid-Mississippi Valley, and along most of the Atlantic coast. These indications can be compared with Figure 11–1.

Thus, with the thought in mind that the British results probably give a general indication about hemispheric climatic patterns likely to accompany the first steps toward a "greenhouse world," let us consider those patterns.

Maximum warming is shown in high latitudes and in continental interiors, with mean annual temperature rises in excess of 3.5°F indicated in northwest Canada, and from Finland across northernmost Russia and Siberia to about 120° E. Except as previously noted, all of North America is indicated to be warmer, as is most of Europe and Asia. There are several regions of minor cooling suggested, however. These regions include much of southern Mexico, Spain and Portugal and the adjacent areas of North Africa, Turkey and other regions around the Black Sea, northeast Libya, northern Egypt, Israel, Iraq, northern Saudi Arabia, much of India, much of southeast Asia, South Korea and Japan, and an area eastward from the Caspian Sea through southern Russia across extreme northwest China into western Mongolia.

Drying is spelled out not only for much of the United States, but for northern Mexico as well. Other important decreases in mean yearly precipitation are delineated for most of Europe (especially France and Spain) and most of Russia, particularly the central Russian plain. Additionally, drying is indicated in Turkey, across northern Africa west of 20° E, from western China into much of southeast Asia, and for Japan and Korea.

Increased rainfall is suggested for other regions of the Northern Hemisphere: Alaska and much of Canada, most of Norway and Sweden, most of Yugoslavia and Greece, a large area of the Middle East, northern Africa east of 20° E, northwest Saudi Arabia, much of Pakistan and India, and most of eastern China and Mongolia.

Of particular interest in the British work is the indication that much of Europe and western Russia share in a combination of warming and drying. As in the United States, such a climatic trend would have serious, negative implications for agriculture. It is easy to dismiss the greenhouse threat as a buffer between our current temperate climate and an ice age, or to shrug it off as merely a hypothetical threat to be poked and prodded by academics. But the fact is, in terms of humankind's ability to feed the world, the threat is real and it is enormous. We may face no more important environmental challenge in our lifetime.

12 THE ULTIMATE POLLUTANT

Assuming that we do not deviate from our fossil fuel course within the next couple of decades, the CO_2-produced climatic warming may begin to accelerate early next century. From a climate similar to that of the 1930s in the period 2010–2020, we would go into to one warmer than anything humankind has experienced in the last 1,000 years. Thus, by 2020–2030 the hemisphere would be 2 to 3°F warmer than it is now. To find an acceptable analog for such warmth we have to turn to a time known as the Medieval Warm Period, or secondary climatic optimum (the primary climatic optimum occurred circa 5,000 BC to 3,000 BC, and was the first major climatic epoch after the last ice age), which prevailed between about 800 and 1200 AD. Research climatologists estimate that the hemispheric temperature during those mild Middle Ages was almost 2°F higher than at present.

ENGLISH WINES

As noted in Chapter 3, the Medieval Warm Period was an era of exceptionally good weather in Europe. Vegetation and glacier boundaries were 500 to 600 feet higher than today, grapes were grown in England and East Prussia, and sheep and woodlands were found in southwest Greenland.

The vineyards in western and middle Europe extended some 250 to 300 miles farther north and up to 600 feet higher above sea level than they do now. This suggests that summer temperatures there averaged nearly 2°F warmer than today.[1]

The mild summers of the time encouraged the growth of the wine industry in England, where current May or July temperatures (or both) preclude such ventures, at least on a commercial scale. William the Conqueror took a census of all landowners and their possessions in his newly conquered kingdom in 1085. The census lists thirty-eight vineyards in England, besides those of the king. They ranged in size up to ten acres, and five of them operated for more than a century. It was said (by an Englishman, I presume) that the wines were comparable in quality to those of France.[2]

In the oceans, Arctic pack ice melted so far back that appearances of drift ice in the waters near Iceland became virtually unknown between 1020 and 1200. Permanent ice probably was limited to the inner Arctic areas north of 80°N.[3]

The northern reaches of the Atlantic Ocean became relatively storm-free. Celtic missionaries from Ireland were able to journey as far as Africa and Iceland. The Vikings settled Iceland and parts of Greenland, and sailed on timber-gathering voyages as far as Labrador.

Eric the Red, banished from Iceland in 982 for killing two men, became the explorer who led the way to Greenland. He settled near a deep fjord on the southwestern coast, and started a Norse colony that at one time had perhaps 3,000 people and 280 farms.

It was at best poor farm country, but the Norse were able to grow vegetables and hay, and live mainly on livestock farming. Eric's own homestead, uncovered by expeditions within the last 100 years, boasted four cowsheds with forty stalls. Today that region of Greenland is barren tundra, and many of the fjords are blocked by advancing glaciers.[4] Evidence from Norse burial sites and plant roots in ground now permanently frozen indicates that mean annual temperatures in southwest Greenland 1,000 years ago must have been 4 to 7°F above present temperatures.[5]

But, as our examination of the 1930s showed, during a time of overall warming mild summers can be occasionally offset by bitter winters. So it was in the Middle Ages. During the era in which the Vikings settled Greenland there was a record of unusually harsh frosts in the Mediterranean, the Tiber River in Rome and even the

Nile in Cairo froze over once or twice, and winters in western Europe were often on the severe end of the scale.[6]

A 200-YEAR DROUGHT

Precipitation patterns, as well as temperature regimes, changed during the Medieval Warm Period. For instance, a bridge built over a river in Palermo, Sicily, in 1113 was designed to span a stream much wider than the one that flows there today, suggesting that Mediterranean rainfall was markedly greater then than now.[7] Researchers have also determined that the period was probably wet on the Yucatan Peninsula, in Cambodia, and in the Near East. There is evidence that the Sahara, too, may have been moister.[8]

But what about the United States? Are there signs that might indicate a 1930s-type drought would be only a passing phenomenon, a transitory stage of the greenhouse effect? Unfortunately the answer is no. Archaelogical studies conducted in the upper Mississippi Valley—roughly in the vicinity of Minneapolis, Minnesota—suggest that the period around the year 1000 was warm and dry in that area.[9] Other studies tell a more ominous story.

Long before Europeans discovered America, Indian nations thrived on the land. In the Midwest, one such group of Native Americans comprised a culture archaeologists call Mill Creek, situated in what is now northwest Iowa. The Mill Creek people were farmers who migrated there about the year 900. They grew corn in the valleys, and hunted deer in the tall grasses and scattered trees of the prairie. But the settlement did not survive. The bones of the game the Indians fed on, and 1000-year old grass and tree pollen trapped in the Iowa soil tell us why.

The bones indicate that the deer and elk, which made up virtually all of the Mill Creek people's diet to begin with, started to disappear. During the 1100s bison became the Mill Creek meat staple. Apparently the tree-browsing deer and elk, which thrive when good rains favor trees, wandered elsewhere in search of food, and were replaced by bison, which graze on dry grasslands. The pollen records confirm a dwindling number of trees and an increase in short prairie grasses during the 1100s. The reason: an unprecedented drought, one that was to last 200 years! After 1300 the Mill Creek villages wasted away. Dr. Reid Bryson of the University of Wisconsin points out, "Clearly two hundred years of drought in the 'breadbasket' of North America is possible."[10]

Thus, our analogs—the warm 1930s and Middle Ages—suggest more than a relatively brief dry spell is in store for the midwestern United States. The analogs squeezed into the 2010 to 2030 time frame warn of fifteen to twenty years of severe desiccation and record-breaking heat. Further, climate studies indicate that the entire eastern two-thirds of North America was warm during the Medieval Warm Period.[11] So the blistering plains heat may sweep to the eastern seaboard in the third decade of next century, engulfing our most populous regions.

And ever so slowly—we would not even notice it at first—the sea level would begin to rise. Thin layers of polar ice would start melting.

THE WEST ANTARCTIC ICE SHEET

The warming would accelerate. Within the next decade, by the year 2040, our climatic temperature may soar to beyond anything known within the past 125,000 years. The average hemispheric temperature could be as much as 5°F above what it is now. It would be a period equivalent to the interglacial that occurred between the last two major ice ages, the Wisconsin of some 20,000 to 60,000 years ago, and the Illinoian, which began about 130,000 years ago.

We can guess, although we do not know for certain, that the great midcontinent drought would persist, perhaps centering over the northern United States and southern Canada. The heat would grow even fiercer; summertime temperatures on the Great Plains might frequently reach or exceed 120°F. U.S. agriculture would be struggling to survive, frantically trying to adjust to vastly altered patterns of temperature and rainfall. Food prices would soar and rationing might become necessary.

Meanwhile, in the Southern Hemisphere, the West Antarctic Ice Sheet would begin to disintegrate—remember the climatic warming is magnified in the polar regions. The West Antarctic Ice Sheet is the smaller of the two ice masses that cover Antarctica, and is a marine ice sheet grounded as much as 8,000 feet below sea level. The eastern sheet is larger and mainly land-based, and would therefore take much longer to melt than the western ice block.

The destruction of the western ice sheet would not be immediate, but within a matter of several hundred years the sea level around the world would rise by an average of sixteen or seventeen feet.[12] Other permanent ice and snow melt would contribute to the elevating seas, and the total rise could be as much as sixty feet! (During the inter-glacial between the Wisconsin and Illinoian Ice Ages the oceans are estimated to have been twenty-five to sixty feet higher than at present.)[13] We would lose our coastal cities. Of course the sixty-foot figure assumes that the climatic warming stops at that point. In all probability it would not, and the oceans would continue to claim the world's coastal land masses.

The melting of the East Antarctic Ice Sheet alone would add another 165 feet of water to the earth. (The western sheet would add only sixteen or seventeen feet because it is smaller, and because it is largely sea ice—which only displaces ocean water.)

While neither we nor our children will live to see the oceans rise significantly, we are the generations who will have to make the decision as to whether, for instance, we risk sacrificing most of Florida to the Atlantic Ocean and Gulf of Mexico in centuries to come. Because once the greenhouse threat becomes reality there is no turning back the inexorable ice melt and the eventual takeover of our coastal homes by the seas.

ALASKAN CAMELS

We might not have to rely on 125,000-year-old indications to get a feel for what the CO_2-fostered climate of 2030 to 2040 may be like on a global scale. A climatic warming of 4 or 5°F would thrust the world into a climatic environment similar to one that prevailed much more recently, about 4,000 to 8,000 years ago. That was an era called the Post-Glacial Optimum, or the Altithermal Period. Figure 12–1 gives a rough idea of the precipitation regimes that accompanied the period, although the regimes did not necessarily occur simultaneously. Evidence for the precipitation patterns and attendant temperature regimes was gathered from exhaustive studies of marine and animal fossils, vegetation and bog growth patterns, tree growth

Figure 12-1. An estimate of the precipitation regimes that accompanied the Post-Glacial Optimum, which occurred roughly 4,000 to 8,000 years ago. A climatic warming of 4° or 5°F—which may take place by the year 2040—would thrust the world into a similar climatic environment. (Map by Dr. W.W. Kellogg, National Center for Atmospheric Research, Boulder, Colorado, reproduced with permission from Dr. W.W. Kellogg. In Gribbin, J. "Fossil Fuel: Future Shock?" *New Scientist*, 24 August 1978, p. 541.)

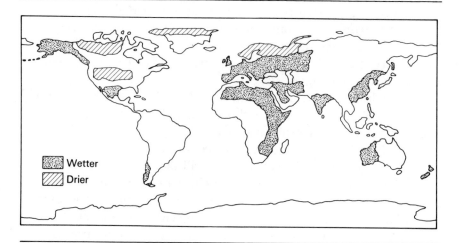

Wetter
Drier

rings, ice cores, and pollen buried in sediments—climatic records etched in nature's log book. That book tells us much about the climates of 4,000 to 8,000 years ago.

In northern latitudes, the Arctic Ocean then was probably ice-free during the summers, but not year round. Camel remains found in Alaska and tiger fossils uncovered in the New Siberian Islands at approximately 75°N suggest that the contemporary climate in those regions was at least temperate (although probably not tropical.)

In Europe, summer temperatures are indicated to have averaged 4 or 5°F higher than at present, with annual means running 3 or 4°F over modern normals. Additionally, Europe was thought to have been generally wetter, possibly due to an enlarged Baltic Sea. (Oceans, near the end of the Altithermal Period, were probably about ten to fifteen feet higher than they are currently.) To the north, Scandinavia was most likely drier.

Vegetation belts were displaced northward and to greater heights above sea level than is now the case. The snow line in central Europe was about 1,000 feet above present elevations.

The drier weather of Scandinavia extended into Russia, and the parched steppe landscape of eastern Russia reached westward into Leningrad and the Volga River Basin. But at the same time much of the great grain-producing region of southwestern Russia shared in the moister conditions of Europe.

This was not true in the American granary, however, where there was a belt of dry grasslands. Temperatures in North America were somewhat warmer than at present, but probably not as much above current levels as they were in Europe. In the United States, both Alaska and Hawaii were wetter than in modern times.

Other areas benefiting from increased precipitation were North Africa (the Sahara) and the deserts of the Near East, thanks to more widespread summer monsoon rains. More abundant winter rains may have blessed the Mediterranean area.

In the Southern Hemisphere, in the temperate middle latitudes, the climate was somewhat wetter and warmer than now, similar to the European changes. Western Australia, New Zealand, and eastern Africa all had more rainfall than currently, but temperature anomalies were probably somewhat smaller than those of Europe.[14]

THE AGE OF REPTILES

Let us suppose that through an active decision to allow the continued use of fossil fuels (or a passive one—no action at all), the full magnitude of the greenhouse threat becomes actual. Where might we be, climatically speaking, in a little over a hundred years from now, or near the beginning of the twenty-second century? In a geological time sense we would probably be near the end of the Mesozoic Era, the "Age of Reptiles," which occurred about two million years ago. The termination of the Mesozoic Era is known as the "Time of Great Dying." It was a period when the dinosaurs disappeared from earth and many small marine organisms became extinct. Their demise was brought about in large part by a relatively quick and significant climatic warming, a warming intense enough to melt the ice covering the Arctic Ocean.[15]

Several researchers have studied the implications of an ice-free Arctic Ocean, a condition that would result if hemispheric temperatures averaged about 10 or 11°F higher than at present.

In such a circumstance, the winter westerlies would probably blow in a pattern very similar to the current May or September pattern.[16] Winter rains would most likely cease in the southwestern United States from California eastward through Nevada, Utah, Arizona, and New Mexico. Herman Flohn, a German climatologist at the Meteorologisches Institut der Universität Bonn, feels that there would be "catastrophic consequences for the water supply of California and Utah." Similar severe problems would crop up across the Mediterranean and in the Near East to Pakistan and Russian Central Asia.[17]

Regional drought conditions at and south of the equator would intensify in northeastern Brazil, in Africa near the mouth of the Congo River on the west coast, and in various east central parts of the continent.[18]

Summer droughts would become frequent between 45 and 50°N.[19] Dust bowl-type weather would settle in as a relatively permanent feature from eastern Montana through the Dakotas into Minnesota and across the southern Canadian prairies. Similar conditions would likely stalk the great Russian granary regions from the Ukranian Soviet Socialist Republic eastward. The United States and Russia currently account for roughly a quarter of all the grain produced in the world. Agricultural practices and regional cultivations may have changed significantly by the end of another century, but one can still suspect that a huge portion of the world's food supply could fall victim to the greenhouse threat by early in the twenty-second century.

The rise of the oceans would have accelerated a bit by then, and sea levels could be three or four feet above what they are now. In many areas that would not be noticeable except during storm tides, but for some cities, such as Venice, Italy, it would spell doom. Low-lying lands such as the Netherlands, Bangladesh, and Florida would be on the verge of inundation.

The oceans would be much warmer and fish populations would migrate toward the poles. Ocean vessels would be able to cruise on ice-free great circle routes over the North Pole. But while warming would dominate most of the world, that might not be the case in certain localized regions.

The British Meteorological Office ran a computer model of atmospheric circulations in which the Arctic ice was assumed to have melted. An unexpected result was that significant cooling—up to 14°F—was forecast in middle latitudes, particularly over the United States, eastern Siberia and western Europe.[20] This could be the result of a problem with the model, but at the very least it points out the fact that hemispheric warming can never be expected to be uniform, a point discussed earlier. It might be a correct prediction, too, a reflection of the severe winters that seemed to have complemented the climatic warmings of the 1930s and the Middle Ages.

CROCODILES IN NEW YORK

Eventually the climate of the entire earth would become tropical, if not unbearable. Over the course of several thousand years the oceans would rise to levels as much as 250 feet over what they are now. Entire states and countries would disappear under water. Many of the world's major cities—New York, Los Angeles, London, Tokyo, just to name a few—would cease to exist. Figure 12−2 indicates the consequences of a 250-foot rise in sea level for the eastern United States. At the 250-foot level the Statue of Liberty would be submerged except for her head and upraised arm. A wide estuary would extend up the Hudson River to Albany, and up the Mississippi River Valley to southern Missouri. The Central Valley of California would be an inland sea. All together about 20 percent of the earth's present land area would be lost; however, there would be a gain of about 10 percent from land uncovered by melting ice and snow.[21]

The nightmare qualities of runaway climatic warming are reflected upon by Lawrence Pick, a fictional character in Arthur Herzog's popular novel about the greenhouse effect, *Heat*:

> He . . . let his mind run. It came to New York years from then if the heat rise continued. Greatly shrunken by the rising sea, the city would resemble a huge Mayan ruin, with the tall buildings covered with creepers and moss. Perhaps crocodiles would float in the lakes and reservoirs, wild animals graze on streets carpeted with grass, vultures fly overhead. Around the city would be jungle inhabited by small bands of humans gradually reverting to savagery in suffocating heat that made civilized life impossible. Maybe hunting parties would come to the island in canoes over rivers swollen by the ocean, in search of food.[22]

Figure 12–2. The consequences of a 250-foot rise in sea level for the eastern United States. A wide estuary would extend up the Hudson River to Albany, and up the Mississippi River Valley to southern Missouri. If all of the world's ice and snow were to melt as a result of climatic warming, the sea level would rise about 250 feet over the course of several thousand years. (*Ice Age Lost*, Anchor/Doubleday, copyright 1974 by Gwen Schultz.)

FIFTY YEARS TO SWITCH FUELS?

Let us become optimists for a moment, though. Let us assume that through unprecedented foresightedness we are able to limit the world's energy growth rate to around 3 percent or a little less over the next few decades. Let us further assume that nonfossil energy sources predominate shortly after the turn of the century. The end result probably would be a climatic warming of slightly over 2°F by the year 2050. (See Chapter 2.)

The ensuing climate would be similar to the one that dominated during the Medieval Warm Period, and the midwestern United States would still be threatened with crippling drought and heat. But the analysis of "market penetration time"—the time required by a new energy source to increase its share of total energy production from 1 to 50 percent—suggests that it may take at least fifty years to replace fossil fuels with nonfossil fuels.[23]

The minimum time frame of fifty years has been determined by analyzing the number of years it took us to change from wood to coal and from coal to oil. This schedule may not in fact be applicable within the framework of modern economics and technology, but it at least tells us there is a substantial period of time required to move from one major energy source to another. Thus, even the optimistic scenario has no chance unless we begin acting swiftly. We can ill afford to sit around waiting for "technology to save us." The technology is here! But a determined and immediate political, social, and economic effort is required to bring the strength of that technology to bear on the problem. The key is immediacy. If we procrastinate for another ten or twenty years, technology may not be able to bail us out. In that circumstance the greenhouse threat could well control our destiny.

WINNERS AND LOSERS

In consideration of our destiny, let us take a closer look at the relationship of the greenhouse threat to agriculture. Specifically, let us imagine ourselves fifty years from now entering an Altithermal-type climate, the climate regime that prevailed 4,000 to 8,000 years ago.

The Altithermal analog—as well as all the other "real earth" models of warm climates we have looked at—suggests immense water supply problems in the western United States, particularly from the Great Basin into the Great Plains. That factor in itself could stifle western farming. But beyond U.S. water supply, the changed precipitation and temperature patterns of the Altithermal Period have significant and far-reaching implications for the future of agriculture on a global scale.

The earth's population is currently a little over four billion people. By the middle of the next century at least ten billion inhabitants will be competing for food, shelter, and energy. And the population could be as high as sixteen billion by then if we revert to the growth rate of 1.9 percent per year that prevailed in 1970. Even if the population grows at the 1975 rate of 1.64 percent, we will have about three times as many mouths to feed within another seventy years!

Record crops in recent years, thanks largely to good weather, have made us forget how dangerously low our food reserves were in the mid-1970s. In 1976, world grain reserves had dwindled to just thirty-one days (based on our annual grain consumption rate.)[24] Dr. James McQuigg, former director of the U.S. Center for Climatic and Environmental Assessment, points out that the difference between a good harvest year, and a bad one (such as 1972 or 1974) amounts to about 10 percent, or 36 days, of annual consumption.[25]

Good years and bad years aside, we are probably in trouble anyhow. John H. Sullivan, a high-ranking official in the State Department's Agency for International Development (AID), warns that not even productive American farmers "can produce enough food to feed all the mouths in the developing world." A food shortfall of 60 million metric tons is foreseen by 1985.[26]

Sullivan thinks the world has been lulled into complacency by a succession of good harvests in India, a major Asian grain producer. He predicts that by the mid-1980s severe famine will strike parts of Asia, resulting in widespread starvation and in a continuation (or resumption) of high food costs and inflation in the United States.

The longer range picture may actually be slightly brighter for the world as a whole, and slightly bleaker for the United States. On the bright side, there is a lot of cultivable land left in the world. The total amount of land on the surface of the earth is 32 billion acres, of which about 8 billion acres could be put to productive use. Currently, only 2.7 to 3.5 billion acres are under cultivation.[27] That leaves over 4 billion additional acres that could be used to grow food.

Dr. Roger Revelle, a University of California professor of science and public policy, speaks optimistically of some of the land that could be put to use. "The southern Sudan, the great plains of Central Africa, could grow enough food to feed the present world population all by itself. And the Ganges plains of India turn out to be one of the most fertile regions on earth, capable of producing as much food . . . as all the U.S. wheat lands."[28]

Under the Altithermal Period climate scenario, both the Sudan (in northeastern Africa) and India would be "winners" in terms of precipitation: they would get more of it. That makes them even more attractive as major world breadbaskets.

Another winner would be Russia. While any increase in temperature would have a negative effect on spring and winter wheat yields, increasing precipitation in the Soviet granary would likely act as an offsetting factor. If the Russians were able to divert water from northward flowing rivers to new lands, they could double their grain production.[29] Clearly, a limited greenhouse threat is not so frightening for the U.S.S.R.

On the losers side of the ledger, chalk up the United States. U.S. grain production would likely decline under an Altithermal-type regime. It is possible that the application of high technology could ameliorate somewhat the effects of diminished precipitation accompanying such a climate, but the cost would be enormous. In combination with the water supply miseries likely with a warming world (such as the lowered water levels of the Ogallala Aquifer) the economic impact on the United States could be staggering.

If the United States were to lose its capacity to export grain, about 21 percent of our current total exports would be wiped out. Our balance of payments scale would be tipped even farther from the black. A point worth noting here is that the biggest single item eating away at our balance of payments now is imported oil—a fossil fuel. Our continued reliance on the fossils, then, may come back to haunt us *ad infinitum*: the fossils leading to anthropogenic climate change, which leads to drought, which leads to diminished crop yields, which lead to diminished exports, which lead to ever larger deficits in our balance of payments.

Even for nations like Russia, the realization of the potential benefits of the greenhouse threat would not come without great cost. One of the natural adversities that would have to be overcome would be that of the inability of soil types adapted to present climates to change as fast as the climate itself.[30] Such problems, in order to be dealt with efficiently, would demand huge inputs of money, labor, and technology.

At the very least, it would seem almost mandatory that—based upon probable CO_2-induced climatic changes—we develop contingency plans for redefining agricultural areas. For instance, in the U.S. we might envision the great wheat growing regions of the Midwest being shifted eastward into the Ohio Valley, or wheat production being expanded in the Pacific Northwest. We might plan that soybeans, a more resilient crop, be grown with greatly expanded acreage in Missouri and Arkansas. Perhaps we should consider replacing tobacco crops with food crops in Kentucky and the Carolinas. Obviously, any decisions on what might be done would extend far beyond the consideration of climatic shifts. They would reach into the economic, political, social, and agronomical arenas as well. But this complexity should not prevent us from planning.

Stephen Schneider of the National Center for Atmospheric Research sums up the world's agriculture future:

> As for the long-term food outlook, there is ample reason for optimism — unless . . . some . . . ecological or political obstacle proves insurmountable. But we should not delude ourselves into thinking that such progress will be cheap or easy — or guaranteed! It will require massive infusions of capital from the rich nations and often dramatic, unpopular changes in existing social, cultural, and political patterns of the poor and the rich.[31]

One thing is certain, none of us will live to see the cost of eating recede. However, by facing the greenhouse threat we can perhaps lower the potential for a dizzying spiral in food prices. Of course, weaning ourselves from fossil fuels would not be cheap, either; but as the television oil filter salesman says, "You can pay me now, or you can pay me later."

"Later" is always more expensive.

13 THE ENERGY OUTLOOK

In view of the acute climatic threat presented by a continued reliance on fossil fuels as our prime energy suppliers, it would be well to look at where the United States appears to be headed in regard to energy sources. The CO_2 problem is global in scale, of course, but the United States, as the world's leading energy consumer, could blaze the trail toward reliance upon nonfossil, renewable energy.

One option we have, naturally, is to do nothing innovative at all. We can continue to employ oil, coal, and natural gas as our dominant energy sources, and to develop synthetic fuels in order to reduce our dependence upon foreign petroleum. In choosing this option we would have to hope that the scientists are wrong—that the greenhouse threat is not serious, at least not on the time or climatic scales proposed. Certainly we would want to have contingency plans ready under this option to meet the threat if it did materialize.

GROWING MORE TREES

Assuming that in trying to exercise this option we suffer the consequences of anthropogenic climate change, is there any hope of reversing the effects? In a practical sense, no, at least not quickly enough to make any difference. If we were to cease burning carbon-

141

based fuels once we realized the damage we had done, natural processes would require centuries to remove the excess CO_2 from the atmosphere. The precise turnaround time would depend upon how quickly the oceans could take up the CO_2 (which is determined by the cooling rate of the seas, changes in salinity and acidity, and marine biotic processes.)[1]

Any buildup of CO_2 could be stripped chemically from the atmosphere, too. That could remove unwanted CO_2 from the air much faster than natural processes, but the cost would be staggering. A back-of-the-envelope calculation indicates that it would take 1,000 chemical plants, each over 300 feet high and 0.6 mile long, operating for thirty years to remove excess CO_2 from the air should we allow the concentration to double.[2]

In theory, technological solutions to CO_2 buildup could be applied just as well to the front end of the problem as to the rear. That is, carbon dioxide could be taken out of combustion waste gases before they ever get into the atmosphere. Optimists point out that there are no obvious technological problems to removing CO_2 from the gases produced by burning fossil fuels, but Dr. Ralph Rotty of the Institute for Energy Analysis, Oak Ridge Associated Universities, disagrees. He describes the scrubbing technique as "very expensive, very difficult, and very energy intensive."[3]

Sulfur dioxide (SO_2)—a contributer to acid rain—is being effectively scrubbed out of the combustion waste gases of many large fossil fuel-burning industries now, but SO_2 is classed officially as an atmospheric pollutant, and the scrubbing has been legislatively mandated. Carbon dioxide is not considered a pollutant—at least not at present—and until it is (and no such movement is under way at the current time) there will be no incentives for industry to develop and employ CO_2 scrubbers.

Other suggestions for dealing with CO_2 have included collecting it before it is released into the air and then injecting it into the ocean. A quick calculation indicates that for a recovery of 90 percent of the CO_2, costs to the consumer should be roughly 20 percent of the fuel costs.[4]

Most of the technical approaches to eliminating excess atmospheric CO_2 have been scoffed at by some scientists as being "really nonsense" because of the large volume of carbon dioxide involved. Dr. Rotty has said the best control measure is "to grow more trees."[5] (Remember that trees and plants use CO_2 in the photosynthesis process.) One can imagine the number of trees required. For instance, one medium-sized one-ton tree might be able to consume effectively the CO_2 produced by burning one barrel of oil.

OUR ENERGY FUTURE

Unquestionably the best solution of all is to quickly diminish our reliance on fossil fuels and remove the problem before it has a chance to overcome us. Is the United States headed in that direction? Table 13−1 summarizes our energy sources as of 1977. Over 90 percent of our energy came from fossil fuels. As for the future, consider one of the most conservative and probably most realistic energy resource predictions issued recently, that of Earl T. Hayes, former chief scientist at the U.S. Bureau of Mines, and now a private consultant. His forecast is based on the predication that our annual energy growth rate will slow from 3 or 3.5 percent to less than 1 percent by the year 2000. The decrease, he feels, will be occasioned by involuntary conservation brought on by higher energy costs and decreased supplies.[6] (Although Hayes thinks a growth rate of less than 1 percent portends disaster because energy and gross national product growth rates have gone hand-in-hand for almost forty years, there are strong indications that this may not be the case any longer. In 1977 the American economy expanded by almost 4 percent while the energy growth rate was held to less than 2 percent![7] So, even before the energy crunch of the late-1970s, we were apparently learning to use energy more efficiently.)

Table 13−1.   U.S. Energy Sources−1977.

	Percentage of Total
Oil	48
Gas	26
Coal	18
Nuclear	3
Solar[a]	6

[a] Solar is defined broadly here to include all solar-related energy sources including wind, hydro, ocean, biomass, and so forth.

Table 13−2.   U.S. Energy Sources−2000.

	Percentage of Total
Oil	34
Gas	10
Coal	34
Nuclear	12
Solar[a]	10

[a] Solar is defined broadly here to include all solar-related energy sources including wind, hydro, ocean, biomass, and so forth.

Table 13−2 presents Hayes' energy source outlook for the year 2000. Our reliance on the fossils probably will have diminished by then, but they will still supply almost 80 percent of our energy, and the total amount of fossil fuel consumed each year actually will rise slightly. Thus, while we will have curtailed the exponential growth rate of fossil fuel use−at least in the United States−we still will be putting just as much CO_2 into the atmosphere as we are today.

That is hardly encouraging when you consider that many of the less developed countries of the world will be significantly increasing their energy usage over the next several decades. And they will certainly opt for the energy sources that remain relatively cheap−the fossils: coal, oil, and natural gas.

In addition, China, with about one-quarter of the earth's population, is striving determinedly to modernize its economy by the year 2000. The modernization will require vast amounts of additional energy, much of which will be drawn from China's extensive reserves of coal.[8] To be honest, the outlook for effectively stemming the growth rate of fossil fuels in the near future is not bright.

OIL

Most of the growth in the rate of fossil fuel consumption will spring from a rapidly increasing use of coal, while the use of oil most likely will tail off by the turn of the century. Proved reserves of U.S. petroleum liquids peaked in 1970 (at 47 billion barrels), then declined to the present level of 37 billion barrels.[9] And even though oil companies drilled more than 48,000 new wells around the country in 1978—nearly double the 1973 figure—production continued to slide. World production probably will peak within the next fifteen years, topping off at 65 million to 90 million barrels per day, depending on whether deliberate curtailment takes place.[10]

To put the accelerating demand for energy into perspective, the Project Interdependence report issued by the U.S. government in 1977 noted: "Half of all the oil that has ever been produced has been taken from the earth in the last 10 years."[11] While some studies have indicated there may be enough recoverable oil, using conventional drilling techniques, to last the world for sixty to ninety years, many experts feel such outlooks are merely academic. For instance, early in 1979, U.S. Deputy Secretary of Energy John O'Leary warned, "Even if enough oil were available in the mid-1980s and beyond, the world economy might be unable to bear the financial burden."[12]

The CIA foresees major world oil shortages as early as 1981 or 1982, as demand outstrips supply. The demand for Saudi Arabian petroleum alone, the CIA estimates, will soar to between 19 million and 23 million barrels daily by 1985. The grimness of that prediction is underscored by the fact the Saudis probably will be producing only 12 million barrels per day by then. Even at that production rate Saudi oil would begin to expire in about fifteen or twenty years.[13]

The United States in particular has put itself in an untenable energy position. Over 40 percent of all the oil we use comes from foreign sources. And 70 percent of that petroleum comes from OPEC (the Organization of Petroleum Exporting Countries), the thirteen-nation cartel that mandates prices and controls production. In 1972, before the Arab oil embargo, our nation's oil bill was just 4 billion dollars; in 1980 the annual tab will be up to 90 billion dollars! But the cost of the imports is probably the least of our problems.

Almost a quarter of the oil the United States imports comes from Algeria, Iran,* and Libya—all strongly anti-Israeli, hostile to the United States, and not at all averse to cutting off our oil supplies. About 15 percent of our imports are shipped from Nigeria, a country whose long-term stability is questionable. And finally, even Saudi Arabia, an established friend of the United States and purveyor of

---

*In November 1979 the United States began a boycott of all oil imports from Iran. This amounted to about 4 or 5 percent of our total foreign and domestic oil supply.

14 percent of our foreign petroleum, may find it difficult to remain either stable or friendly in the cauldron of Middle East politics.

To be perfectly blunt about it, we probably are about to hang ourselves. The United States is standing on the gallows, the noose around its neck, just waiting for someone to yank open the trap door. If the greenhouse threat alone is not enough to trigger an urgent national resolve to kick the fossil habit and develop clean, renewable energy sources, the additional threat of being strung up by an empty gasoline hose should be.

Unfortunately, one of the cleanest sources of energy—in terms of both recognized pollutants and CO_2—will likely diminish in importance as a major energy supplier over the next twenty years. Natural gas, which currently provides about a quarter of our total energy requirements, mostly to industry, probably will be able to furnish only a tenth of our needs in the year 2000. At the same time, one of the dirtiest fuels, coal, will get an expanding role in the energy scenario.

NATURAL GAS

To the consumer, natural gas remains the cheapest traditional source of heat, but the price advantage of gas over oil probably will disappear by the mid-1980s. Under the 1978 Natural Gas Policy Act, domestic gas prices will rise in steps until 1985, at which time prices will be completely decontrolled. Fully 95 percent of the gas consumed in the United States is domestically produced, but production peaked in the early 1970s and has been declining ever since. Over the past decade we have been using 20 trillion cubic feet (tcf) of natural gas each year and finding only 10.

Production peaked between 1971 and 1973 at 22 tcf and dropped to 19 tcf by 1977. Production has dwindled along with proved reserves. In the late 1960s proved reserves in the United States stood at 290 tcf, but by 1977 had diminished to 209. Earl Hayes foresees domestic production shrinking to 9 tcf per year by 2000. He admits this is a conservative estimate, and thinks that with complete deregulation of prices—allowing them to double over those of early 1979—another 3 tcf could be produced each year. Most of this gas would come from unconventional sources—western "tight" sandstones, Appalachian Devonian shales, and Gulf Coast geopressured regions.

Hayes thinks that imported natural gas will account for about 1 tcf per year by the turn of the century, but also fears that this estimate is "generous in light of the predicted decline in reserves and production of world oil and gas starting in the middle 1980s."[14]

COAL

Coal, which currently provides about 18 percent of our total energy, will become more prominent in our energy future. The United States is sitting on enough coal to last us easily through the next century, and probably the next century beyond that. Clearly we are in a position to become the coal equivalent of OPEC. The drawback is that coal use is fraught with problems. Not the least of those problems is the fact that the combustion of coal releases 75 percent more CO_2 to the atmosphere per equivalent unit of energy than does natural gas, and releases up to 30 percent more CO_2 than does oil.[15]

The dark side of coal becomes evident long before it is burned, however. Even without major disasters, more than one hundred people die in deep mine accidents every year. Additionally, in excess of 1 billion dollars in federal money is given annually to victims of black lung disease.[16]

In the West, deep mines are absent, but strip mines disfigure the landscape. Tight regulations on land reclamation have been enacted, but, as a *Boston Globe* editorial points out, ". . . it is hard to predict just what will happen as increasingly large amounts of coal are ripped from the earth."[15]

Once the coal is out of the ground there are great difficulties in getting it to market. With overworked railroad coal cars and disreputable roadbeds, transportation will be an acute bottleneck for the coal industry for years to come unless immense amounts of capital are immediately poured into the nation's railways. Another strike against coal transportation is the portion of the more than 2,000 annual railroad deaths that can be ascribed directly to the shipment of coal. That is not a problem in parts of the West, though. In some remote coal mining reaches of the Dakotas and Wyoming there are no railroads.

Coal slurry pipelines are perhaps an answer to that transportation dilemma, but the coal slurry process demands water, which is anything but plentiful in those regions. Chapter 4 details the huge water supply quandaries facing the West and, in that light, points out the folly of developing energy sources dependent upon coal or shale oil.

Beyond scarring the land and usurping water from agriculture and direct energy production, coal mining leads to a host of additional environmental problems. Among these are the disposal of sludge and other wastes from pollution control equipment, unsightly transmission lines, reduced atmospheric visibility in wilderness areas, and, in Appalachia, acidic rainwater runoff.

The actual combustion of coal fosters problems other than the release of CO_2 to the atmosphere. According to a study performed by scientists at the Brookhaven National Laboratory on Long Island, and the Carnegie-Mellon University in Pittsburgh, there could be as many as 35,000 premature deaths annually—mostly due to respiratory ailments—by 2010 if coal burning in the United States proliferates.[18] The health hazards associated with coal combustion stem from the waste gas release of sulfur dioxide (SO_2) and particulates (small particles of dust and soot) into the atmosphere. Pure sulfur dioxide by itself is only a mild respiratory irritant, but in the air we breathe, SO_2 is mixed with other pollutants—including particulates— and converted to, among other things, sulfuric acid.

The estimate of 35,000 deaths each year by 2010 is based on the assumption that 80 percent of the sulfur emitted by coal-burning industries and utilities will be scrubbed out of the waste gases before they are injected into the air. Current Environmental Protection Agency (EPA) regulations say that 90 percent of the sulfur in high-sulfur eastern bituminous coal, and 70 percent of the sulfur in low-sulfur western lignite coal must be removed from the waste gases. Newly constructed utilities may burn low-sulfur eastern anthracite without employing scrubbers, however. That is roughly equivalent to burning bituminous coal using controls (e.g., scrubbers.)

The myriad of federal regulations governing the coal industry, though necessary, seem likely to impede—if not stifle—expansion of coal production. As National Coal Association President Carl Bagge put it in a letter to President Carter: "It has become increasingly clear that our national goals are in conflict and cannot all be achieved simultaneously."[19] The government regulations are translated into cost, of course. While a coal-fired utility can be built for about two-thirds the cost of a nuclear plant, coal burners are markedly more expensive to operate. In 1978 the average total costs of producing electricity up to the point of transmission to the consumer were: 1.5

cents per kilowatt hour for nuclear, 2.3 cents for coal, and 4 cents for oil.

Finally, for those who are strongly antinuclear, consider the following nonnuclear "danger": a utility plant burning western lignite coal does not meet Nuclear Regulatory Commission standards in regard to the amount of low-level radiation emitted. Such a plant exposes people to more radiation, on the average, than they would get from a nuclear facility generating an equal amount of electricity. About fifty-three coal-fired plants are scheduled to rise in the West— the home of lignite coal—by 1995.

Despite our country's vast store of coal, its future as a major energy producer—given the array of environmental, health, safety, and economic problems surrounding it—is highly dubious. Until recently it appeared as though nuclear power might be able to take up any slack left by coal. But along with a lot of radioactive water, that possibility went down the drain at Three Mile Island (even though the water was contained and not released into the environment.)

NUCLEAR

Prior to the accident at Three Mile Island, President Nixon's energy planners foresaw nuclear plants supplying 40 percent of all U.S. electricity by 2000. Subsequently President Carter's administration projected no more than about 25 percent of our electricity, or a bit less than 8 percent of our total energy, from "nukes" by the end of this century.[20] (Earl Hayes' estimate—Table 13−2—that nuclear power would furnish 12 percent of our total energy by 2000 was made prior to the events at Three Mile Island.)

Nuclear facilities currently provide about 12.5 percent of our electricity, or something less than 4 percent of all our energy. Given the immense amount of social and political opposition now rallied against nuclear power, there is some doubt it will ever be allowed to exceed even those modest figures.

However, to provide 25 percent of our electricity by 2000, close to 400 nuclear plants will have to be in operation. At the end of 1979 there were 72 nukes on line; utilities had construction permits for an additional 94; and another 34 were in the engineering stage.

The resistance to nuclear energy seems highly emotional, but there are serious questions that remain unanswered. Little data are available on the hazards posed by terrorism, little is known about the effects of leaks from waste storage areas, and there is not a reliable estimate of the risk of a major nuclear plant leak.

Still, the consequences of shelving future nuclear development must be considered. The substitute for each 1000-megawatt nuclear plant that is not constructed is a fossil-fired utility that would burn either 51,000 more tons of coal each week, or 8 million additional barrels of oil each year.

Within the near future the choice between nuclear and fossil-generated electricity will come down to an assessment of risk. We will (1) have to accept the known environmental threats, the known menace to our health, and the certain death rate associated with coal, (2) have to be willing to live with the specter of economic blackmail and chronic shortages should we duck our resolve to end our reliance on imported oil, or (3) have to make a commitment to live with—and try to lessen—the risks posed by nuclear power.

Nuclear power, in another form, may eventually provide the United States the answer to its energy dilemma. Traditional nuclear power is derived from nuclear fission, the splitting of atoms. Nuclear fission's dangers are well-recognized. However, another way of deriving nuclear power, nuclear fusion, is a nonradioactive process. The process involves the fusing under extraordinarily high temperatures of two hydrogen isotopes into helium, the same element that powers the sun. One hydrogen isotope, deuterium, is available in seawater, and the other, tritium, can be manufactured.

Nuclear fusion research has been plodding along for about twenty years, and it probably will be another twenty years before a prototype of a commercial fusion reactor is built. Some researchers feel that fusion might contribute a few percent of our total energy by 2020,[21] but others, such as Dr. John M. Dawson of the Physics Department of the University of California at Los Angeles, think "we can be energy self-sufficient" by 2020, thanks to nuclear fusion.[22] That outlook is comforting, but it still leaves us with a forty-year fossil fuel replacement problem.

SOLAR

Rapid development of solar energy could go a long way toward solving that problem, but a significant increase in solar power—here broadly defined to include wind, hydro, ocean, and biomass—is not foreseen by 2000. Solar energy currently contributes about 6 percent of our total energy with most of the contribution coming from hydroelectric power and biomass. Biomass energy is now mostly derived from burning wood or wood wastes. Earl Hayes' estimate is that solar energy will supply about 10 percent of our need by the turn of the century, a reasonable projection in light of present government policies.[23]

In early 1978 President Carter's Council on Environmental Quality (CEQ) concluded that with the proper incentives as much as 25 percent of the nation's energy could be solar by end of the next two decades. In mid-1979 President Carter himself set a national goal of having solar energy produce 20 percent of the nation's power by the year 2000.

However, solar advocate Barry Commoner attacked Carter's solar program as being "window dressing."[24] And other sources criticized it for lacking specific timetables and goals.[25] Certainly the amount of money committed to a "solar bank" to develop solar power—3.5 billion dollars over the next 10 years—is a far cry from the 40 to 50 billion dollars in federal money that is required to bring solar energy into the 20 to 25 percent contribution range by 2000.[26]

If we are fortunate enough to reach the 20 percent plateau by the end of the century—thanks to visionary leadership—individual constituents of solar energy probably will be able to account for the following shares of our total energy supply: solar heating and cooling, 6 percent; biomass, 6 percent; hydroelectricity, 4 percent; and wind and ocean, 4 percent.[27]

Even by the year 2000 most of the sun's captured direct energy output will be used to heat homes and buildings and provide hot water, as opposed to being used to produce electricity. The making of electricity from sunshine is known as photovoltaic production, a process that converts sunlight directly into electric power by using silicon crystal cells. The current cost of photovoltaic production is prohibitive: about 9 dollars per watt.[28]

Still, the cost of photovoltaic production has dropped dramatically since 1975 when it was 22 dollars per watt. (Efficiency of the cells has improved markedly, too, from about 2 percent in 1969, to as much as 16 percent today.) The Department of Energy has set a goal of further reducing the cost to 2.80 dollars by 1982, 70 cents by 1986, and to no more than 40 cents by 1990. Homes of the twenty-first century will undoubtedly have both solar heating and photovoltaic systems.

BIOMASS

Another source of energy from the sun is biomass—organic matter such as crop and forestry residues, marine plant life, and municipal and animal wastes. (Because biomass depends upon sunlight and photosynthesis, it is considered to be a solar-related energy source.)

Only about 1 percent of our total energy is currently supplied by biomass, but close to 6 percent could be by the end of the century. Burning wood is by far the most promising of the biomass conversions, but a number of other processes show commercial potential. Gasification and liquefaction of wood and low-moisture plants can produce natural gaslike and petroleumlike products. Anaerobic digestion, or treating manure with certain bacteria, can make synthetic gas. And fermentation of wood and plants containing low moisture can produce gasoline substitutes.

Another technique already employed on a commercial basis is the burning of municipal refuse—garbage—to manufacture electricity. At least sixteen plants around the country are currently doing that, including facilities in New York, Chicago, Sacramento, and Milwaukee.

The biomass conversion processes, while releasing CO_2 to the atmosphere, do not add to the total amount of CO_2. This is because the CO_2 released by biomass is already part of the ongoing carbon cycle in the environment. Carbon dioxide injected from fossilized sources is not, and thus adds to the overall atmospheric CO_2 concentration.

HYDRO

Fifty years ago, the United States got over 30 percent of its electricity from water power. But the era of cheap oil saw many dams abandoned or destroyed, and hydroelectricity now accounts for only about 15 percent of our total generation, most of that in the Pacific Northwest where an amazing 80 percent of the region's electricity is furnished by swift waters.

The majority of the commercially promising sites for big dams— sites with steep river drops or wide, fast flows—have been built upon already. The enthusiasm now is for "low head" dams (dams less than 65 feet high). A study commissioned for the U.S. Department of Energy in 1978 indicated there were 50,000 to 80,000 small dams scattered around the country, not generating power. If powerhouses were installed at but 5,000 of these dams, the total megawatt output could exceed the potential of twenty-two large nuclear facilities.[29]

The greatest promise is in New England, where about 2,000 small dams have the latent ability to make electricity. If all 2,000 sites were developed, enough power could be generated to equal the productivity of two nuclear plants. That is an important consideration for a region 63 percent dependent on foreign petroleum.

The biggest problem associated with converting existing dams to hydroelectric facilities is financial. Banks and private investors are reluctant to take risks on small ventures. But, there are environmental problems, too. Fish ladders must be constructed so that migratory fish can safely pass dam sites; there are fears that blocked river flows will silt up river beds; and there is concern that the backed-up waters behind dams will create reservoirs that submerge good land. Still, the optimistic signals are that hydroelectricity will expand slightly from supplying 3 percent of our total energy, to supplying about 4 percent by the year 2000.

WIND

The combination of wind- and ocean-derived power could supply an additional 4 percent by then. But some researchers argue that this is a short-sighted goal. Professor William E. Heronemus of the University of Massachusetts, Amherst, an outspoken advocate of wind power, points out, "Wind alone over the continental United States could generate every bit of electricity this country wants and economically, too."[30]

Dr. M. R. Gustavson of the Lawrence Livermore Laboratory at the University of California agrees with Heronemus on the potential of wind power. Gustavson feels that about 75 percent of our total current energy consumption could eventually come from wind.[31]

Wind energy is traditionally viewed as being especially practical in remote areas where large quantities of electricity are not needed, and where transmission costs from distant utilities are steep. In the United States, wind power potential, based on climatology, is high on the Great Plains, in the northern Rockies, along the Texas Gulf Coast, around the eastern Great Lakes, along the Pacific coast, over Cape Cod, and in the Aleutians. Peak potential is off the Oregon and Washington coasts, over southeastern Wyoming, in the Texas-Oklahoma panhandle areas, and in the Atlantic Ocean off of New England and the Middle Atlantic states.[32]

Professor Heronemus, not one to limit himself to traditional thinking, envisions as many as 40,000 wind turbines rigged in multiple arrays someday floating off Cape Cod, and providing all of New England's electric needs![33]

Wind-generated electricity is not currently cost competitive with that produced by oil, coal, or nuclear plants, but there seems to be little doubt that it will be once wind turbines are mass produced.

OCEAN

The world's oceans can provide us with energy in several ways. Electricity from turbines driven by the power of ocean currents, tides, or waves is an obvious source. But energy can also be derived through Ocean Thermal Energy Conversion (OTEC). OTEC takes advantage of the heat differential found in tropical oceans—warm water at the surface and much colder water underneath. The process requires construction of a floating power plant that uses the warm water to heat ammonia until it expands into a gas and drives turbines that generate electricity. The cold water, pumped up from the ocean floor, then cools the gas to a liquid, and the cycle is repeated endlessly as the ammonia circulates through the system.

Heronemus thinks the potential of OTEC is enormous, even greater than that of wind. He estimates that by using only one two-thousandth of the thermal resource in the Gulf Stream, enough electric power could be turned out to satisfy the entire nation.[34] OTEC technology was tested as early as the 1930s and is known to work, but it remains a very uneconomical source of energy. Still, because of its awesome potential, it deserves more attention than it has been officially receiving.

GEOTHERMAL

Geothermal energy is another source of potentially cheap and relatively pollution-free power that has received minimal attention. Geothermal energy exists in volcanoes, geysers, and hot springs. It can also be tapped by sinking wells roughly 2,000 feet down into underground lakes of superheated water and steam sandwiched between layers of hot rock near molten lava. The steam is brought to the surface where it can be used to turn electricity-producing turbines.

Iceland already extracts much of its energy from the earth's steamy interior, and California currently generates about 2 percent of its electricity from such sources. Union Oil estimates that within another ten years as much as 25 percent of the Golden State's electric power could come from geothermal energy.[35] The hottest prospects for the commercial development of geothermal energy are in the West, particularly along the string of active volcanoes in northern California, Oregon, and Washington. Hawaii, created by volcanic activity, also shows promise of being able to tap the earth's boiling underground resources.

But there are problems with geothermal energy that most likely will retard its quick development. To begin with, it costs more than 500,000 dollars to drill a geothermal well, compared to between 250,000 and 500,000 dollars for a gas or oil well.[36] And the hot reservoirs can be just as difficult to find as oil deposits. Once a superheated lake is found, its steam may have a relatively low temperature that is not particularly efficient for turning turbines. There is also a pollution problem. The steam may carry salts and minerals to the surface that destroy soil in some regions. But this difficulty can be overcome by injecting the hot water back into its reservoir (which might also help to keep the pressure high).

Despite the drawbacks, geothermal, like OTEC, deserves a much harder look. It is an energy source that is virtually inexhaustible, free (once the well is dug), and relatively harmless to the environment.

HYDROGEN

The lightest of the elements, hydrogen, may someday prove to be a heavyweight in the energy ring. Hydrogen can be burned in place of oil, coal, or natural gas. It is a clean energy source, easier to store than solar energy, and so abundant that it could fuel the world. Hydrogen can be obtained by splitting atoms of water, a substance that covers 70 percent of the earth's surface. It can also be stripped out of coal, natural gas, or oil.

The trouble with hydrogen is that it is expensive to process. Currently, for the small user, it costs from ten to thirty times as much as natural gas per equivalent unit of energy. However, some European extraction processes have been able to get the multiplier down to about three.[37] There are myriad other problems that must be overcome before hydrogen can become an integral part of our energy future, not the least of which is the fear of its explosive tendencies. (Hydrogen advocates refer to this as the "H−2 Hindenburg Syndrome.") Still, in the distant future it may be possible to have both nuclear and solar plants churning out enough hydrogen to power everything from jet aircraft to our homes.

THE BEST RESOURCE

There is one final resource that we have, and it is the resource with by far the greatest short-term potential of all: conservation. The United States is tremendously wasteful of energy. We have but 6 percent of the earth's population, yet we consume 30 percent of the world's energy. Based on energy consumption per unit of gross national product, we devour twice as much energy as West Germany or Sweden, and more than twice as much as Switzerland or Denmark.[38] Our automobiles burn up 35 to 40 percent of all the oil in the United States each year, and 10 percent of the world's oil. Obviously there is room for a lot of improvement.

Daniel Yergin, a lecturer at Harvard University, Cambridge, Massachusetts, and coauthor of the best-selling book *Energy Future: Report of the Energy Project at the Harvard Business School*, argues that with only minor life-style adjustments and no decline in economic growth, the United States by the end of the century could consume 30 to 40 percent less energy than it does today. Dr. Warren Johnson of San Diego State University, and author of *Muddling Toward Frugality*, is a bit more definitive when he says that a 30 percent per capita cut in energy usage would return us to a 1950s style of living.[39] The point is that a one-third reduction in our national energy consumption is certainly within reach, and we are reaching for it.

New automobiles, which in 1974 averaged 14.4 miles per gallon (mpg) will average 20 mpg in 1980, and—in compliance with the Energy Policy and Conservation Act of 1975−27.5 mpg in 1985.[40] This will make a significant dent in our dependence upon imported oil. It is a goal that must not be postponed.

Industry has made other strides toward lower consumption. Since 1973 industrial use of energy has diminished by 6 percent while at the same time output has been hiked by 12 percent.[41] In the last decade Dow Chemical, one of the three largest users of energy in the country, reduced its consumption by 40 percent per pound of product. This was accomplished with relatively little capital investment. The Gillette Company, a Boston-based multinational corporation, now uses 30 percent less energy at its U.S. facilities than it did just five years ago. Again only a small investment of money was required.[42]

Other significant energy conservation trends have taken place in U.S. homes since 1973. Half of all homeowners have added insulation to their dwellings, the energy efficiency of residential buildings has increased 5 to 10 percent, and home appliances have increased in efficiency by 5 percent. As a result of both industrial and residential efforts, the annual growth rate in electric demand has been halved (from 6 percent to 3 percent.)[43]

Energy conservation is the cheapest, cleanest, most readily available, and most efficient of our energy resources. It can flatten the exponential energy growth curve, and it can buy us the necessary time to develop acceptable alternatives to fossil fuels. It is a resource we can all tap, and individually control. It is the cornerstone of our energy future. In the sense that soaring energy cost is the arch-villain of inflation, conservation is also the cornerstone of our *economic* future.

14 WHAT CAN WE DO?

Scientists working in the CO_2-climate field stress that the results of their work are preliminary. Still, those results have so far all pointed in the same direction. As Dr. Frank Press, President Carter's science adviser, puts it, "Everyone agrees that CO_2 is increasing and will have a warming effect. But how great the effect will be . . . is not certain."[1] *The Greenhouse Effect* has presented—based on the knowledge available at the end of the 1970s—a sequence of climatic transitions quite possible over the next hundred years or so. True, the scenario is not guaranteed to transpire, but no meteorologist can guarantee his forecast for even the day after tomorrow.

ACTION BEFORE KNOWLEDGE

Researchers feel that more definitive answers to questions about CO_2 and its effects will not be found for another five or ten years; others think it will take at least twenty years. Dr. Roger Revelle is one of the latter: "I don't think we are going to make the assessment that anybody will really believe until we can see an actual rise in temperature which is beyond the noise level [natural fluctuation] of the system. That is liable to take many years."[2] The point to keep in mind, however, is that we do not have a tolerance of "many years"

159

in which to react to the problem. In consideration of the severe climatic consequences that may befall us, the nearly half century required to switch primary energy sources, and the current government program encouraging the development of synthetic fuels that will add even more CO_2 to the air, it is obvious that action to step away from fossil fuels must begin now.

Some scientists have argued quite eloquently for such action, while others are reluctant to suggest what policies we ought to be pursuing. Dr. J. Murray Mitchell, Jr., U.S. Environmental Data Information Service, National Oceanic and Atmospheric Administration (NOAA), is an advocate of action: "Ours is the generation that may have to act, and act courageously, to phase out our accustomed reliance on fossil fuels before we have all the knowledge that we would like to have to feel such action is absolutely necessary. If we harbor any sense of responsibility toward preserving spaceship Earth, and toward the welfare of our progeny, we can scarcely afford to leave the carbon dioxide problem to the next generation."[3]

Other researchers, such as Dr. Kirby J. Hanson, director of NOAA's Geophysical Monitoring for Climatic Change, hesitate to speak out on the matter. Hanson feels that scientists should only advise policymakers on the probability of certain environmental changes happening, and "not attempt to state or make policy."[4] While such an attitude is understandable, it is a little frightening to think that policy eventually would be made by someone without a comprehensive understanding or first hand knowledge of any background research work. In that light, it would seem wiser for scientists to suggest certain policies or courses of action related to their particular disciplines, and then let "official" policymakers evaluate those suggestions in light of the limited fields of view in which they were made.

As an example of a limited field of view, consider the testimony given to a House science subcommittee in September 1979, by Ruth Clusen, Energy Department Assistant Secretary. She stated the interpretation of her staff was that the potential increase in CO_2 from synthetic fuels, by itself, would not be large enough to cause climatic change.[5] Strictly speaking, that probably is true. But it is not the point. The point is we should be advocating policies that carry us *away* from reliance on fossil-based fuels, not suggesting that the development of synfuels is OK because it adds just a small increment of CO_2 to the air. New technologies should be pushed because they are based on nonpolluting, renewable resources, not because their deleterious environmental effects might be below a detectable threshhold. And, as discussed in Chapter 4, the synfuel industry's problems extend well beyond the CO_2 issue. They embrace potential water supply shortages and damage to the terrestrial environment.

AN AMALGAM OF ISSUES

But let us put the entire CO_2-climate problem under a broad spotlight, and consider it in an amalgam of other issues and other goals. While no individual argument is likely to be overwhelming enough to convince everyone of the urgency of ridding ourselves of the fossils, a bank of such arguments may be. In fact, seen in a wider perspective, the folly of our continued dependence on fossil fuels becomes undeniable.

To reiterate, the key issue is the enormous climatic consequences we face when the greenhouse threat becomes reality. The effects may be particularly harsh in the midwestern United States, the heart of American agriculture. We could gamble that the scientists are wrong, and that the effects will not be manifest within the next several generations; but the stakes in this instance are so high as to make such risk-taking foolhardy.

Secondly, there is the obvious need to lower our dependence on foreign petroleum as rapidly as possible. The U.S. economy is at the mercy of OPEC, and we remain wide open to the latent threat of blackmail by countries that control nearly one-tenth of all of our energy resources.

Thirdly, there is the ultimate necessity of developing renewable energy resources anyway. World oil and gas production will peak shortly, and although coal remains in abundance, its environmental and health-related drawbacks are so severe that its future is limited. Synthetics have been touted as an answer to the energy dilemma, but the synfuels face many of the same problems as coal. Additionally, synfuel technology is ten or fifteen years away and the cost ten times greater than a "barrel" of conserved energy—hardly a thrifty answer to our energy crisis.[6]

Finally, there is the consideration that it will take us roughly half a century to switch from one prime energy source to another. Given this inertia, the negatives stacked against the fossils, and the logic in favor of the renewables, it makes sense to begin immediately turning toward replenishable forms of energy. A later change of course would only come amidst greater turmoil and greater sacrifice. A much smoother transition can be made now. The longer we delay the decision, the longer we expose ourselves to environmental, economic, and social threats, and the more vulnerable we become.

Certainly we must discard the hope that technology will somehow magically pull us out of the energy mire we are in. Technology will bail us out eventually, but the process will be neither magical nor painless. The responsibility is on our shoulders to point technology in the right direction and to encourage its development. We can do this through the political process, but we can also contribute directly—and personally. There will be both economic and social sacrifice involved, but the severity of those sacrifices can be mitigated by committing ourselves now to proper energy technologies for the future.

THE UNSOPHISTICATED TECHNOLOGY

The most direct personal contribution toward extracting ourselves from the undesirable energy position we are in can come through conservation. Conservation itself is a technology; an unsophisticated one to be sure, but easily within reach of us all, and very cost-effective. Conservation can reduce our energy costs, lessen our dependence on foreign sources, diminish the amount of CO_2 we are spewing into the air, and buy us the necessary time to develop the renewable energy resources we so desperately need.

It is difficult to accept that the time of cheap, plentiful energy is gone. But the simple fact is that it is. And, in the near term, it is going to get worse. It is easy to single out big oil, and perhaps government, as the villains in the drama. In truth there are not any villains, only the stark reality that energy resources are dwindling and competition for those resources is accelerating. Any finger-pointing should be at our own consumption patterns. We devour a disproportionate share of the world's energy, and not very efficiently.

While it is easier to lay the blame elsewhere, artificially controlling energy prices or breaking up the oil corporations will not help us increase production or lessen demand.

In the end, economic considerations (high energy costs) and not patriotism or philosophy, probably will spur our conservation efforts more than anything else. The economic incentives, which we are already experiencing, will come in the form of continued price rises on the free market, or in the form of government-imposed energy taxes.

The scarcity of supplies, as Earl Hayes argues, will likely dictate a slowing of our overall energy growth rate, and to that extent force stringent conservation upon us. But beyond that, as Daniel Yergin points out, there is room for an actual reduction in the amount of energy we consume. By early next century, assuming continued economic and population growth, we could be using 20 percent less energy than today and enjoying virtually the same standard of living![7] Such a reduction, supported by strong leadership and active government programs,[8] would require only relatively small capital investments, but would have the potential for immense returns.

The steps we can take to make our homes more energy efficient are generally well known now, thanks to efforts of the federal government, state and municipal agencies, and public utilities. And the economic benefits of higher mileage automobiles, car pooling, and mass transit are becoming more apparent to us each day. Our personal conservation awareness is being steadily heightened. But some cities in the western United States have demonstrated that such awareness need not be limited to individual efforts.

"A DAMNED GOOD DECISION"

Portland, Oregon, is implementing what has been called "the boldest, most sweeping energy conservation program in the nation."[9] Among its provisions are ones that:

1. Would give homeowners five years in which to "weatherize" their homes well enough to pass a city energy audit at the end of that five-year period. Homes not rated "energy efficient" would not be permitted to be sold. Similarly, apartments not meeting the city standards could not be leased to new tenants, and businesses would not be allowed to remodel or sell noncomplying buildings.
2. Would revise zoning laws to encourage increased population density near public transit routes.

3. Would authorize the city to hire a full-time solar energy expert, and would require builders to provide clients with information on solar technology for new construction.
4. Would require trash haulers to offer to customers free of charge the option of separating their recyclable garbage; the haulers would then recover their costs by selling the reclaimed materials.

(The mandatory aspects of Portland's plan are meeting strong opposition. They have been attacked as being unconstitutional, and probably will face some serious challenges in the near future.)

In Seattle, Washington, Portland's neighbor to the north, Seattle City Light halted plans to erect a nuclear power plant. It instead launched an energy conservation drive that reduced electric demand growth to more than 2 percent below projections. Peter Henault, environmental affairs director for City Light, explained the utility's choice. "We thought conservation was a lot cheaper, a lot less environmentally harmful, and just made more sense. It's turned out, everyone now agrees, to be a pretty damned good decision."[10]

Seattle is also the first known city in which a major employer has made transit passes available as a permanent, completely cost-free fringe benefit to its employees. Seattle First National Bank offers, to thousands of its workers, free yearly passes on local bus systems.[11]

Seattle's modern bus fleet carries nearly half of all that city's commuters. And the figures are similar for the bus-based mass transit systems of Portland and Denver, Colorado.[12] Buses appear to have an important future in mass transit, and will likely fill the role of "a very major energy saver." (Heavy rail systems may actually consume more energy than they save, mainly because of the energy required first for their construction, and then for getting passengers to stations along the line.)[13]

TAX CREDITS AND GOLF CARTS

In addition to being a leader in conservation, the far West is also in the vanguard of development of our most important renewable energy resource: solar. California, with about one-tenth of the U.S. population, has about one-quarter of the nation's solar applications. California is climatically suited for solar development, of course, but in 1977 the state legislature voted in a 55 percent solar tax credit—the single largest financial incentive in the country for solar energy.[14]

Oregon has offered a 25 percent tax credit to encourage wind, solar heating, and geothermal development since 1975. And Portland General Electric has agreed to buy back, at its own rates, electricity produced by home solar generation. Some Oregonians are building small hydroelectric facilities, while more than a tenth of the state's residents are using wood as their chief fuel source.[15]

A U.S. Department of Energy employee, Duane Day, has noted that the most successful solar energy programs have been in states that offer substantial tax credits.[16] Modesto A. Maidique, writing in *Energy Future*, suggests that tax credits at least as large as those offered by California be applied on a federal level. With such a program, Maidique argues, we could expect solar to fulfill 20 percent of our energy needs by the year 2000.[17]

Additionally, federal financial support to solar research and development must be dramatically increased in order to spur further development of solar technologies, particularly those that promise the most immediate returns—on-site solar applications (heat, hot water), photovoltaics, wind, and biomass energy "plantations." Between 40 and 50 billion dollars in government funding is needed over the next two decades to bring solar energy to the 20 to 25 percent contribution level.[18] That is roughly twice what we spent in the 1960s to put a man on the moon, and about half the 88 billion dollars that President Carter ticketed for the development of synthetic fuels.

The total national investment, including that by industry, venture groups, and individuals, would be considerably more, of course. If, for example, photovoltaics (at the 1986 goal of 70 cents per watt) were to furnish all of our new electrical generating capacity—assuming an annual demand growth rate of 3.5 percent, half the historical rate—up to the end of the century, a total outlay in excess of 1.5 trillion dollars would be required.[19]

It would seem eminently wiser, rather than committing 88 billion dollars to the dubious future of synfuels, to put a healthy share of that money into solar energy, a clean, renewable source. If we can spend 22 billion dollars in a decade to get a golf cart on the moon, it would seem that 45 billion dollars, spread over two decades, is not an unreasonable sum to request for our energy independence and environmental well-being.

We need not demand a Manhattan or Apollo-type crash program to achieve the maturation of solar energy (solar technology is actually too diverse and too diffuse to lend itself to such a concentrated effort), but we must seek for it the same sense of urgency and support that thrust us into the atomic era and launched us into the space age.

GAS AND NUKES

However, conservation and solar energy alone will not tide us over until the end of the century. Conventional sources will still have to be called upon. The domestic production of natural gas—the least harmful of the fossil fuels—should be encouraged to the maximum extent possible. By completely deregulating the price of natural gas we can probably be producing about 13 trillion cubic feet (including 1 tcf imported) per year at the turn of the century.[20] This would furnish about 14 percent of our energy needs, as compared to Earl Hayes' more conservative estimate of 10 percent (See Chapter 13).

In the scenario of limited energy growth and solar power expanding to cover one-fifth of our requirements, our reliance on coal and oil by the year 2000 could be somewhat less than usual estimates suggest. For instance, our dependence on oil could be reduced from Hayes' estimate of 34 percent to about 26 percent, with 60 percent of that petroleum coming from domestic sources. Coal would still have to be heavily exploited, but our reliance on it could be drawn down to about 30 percent (from Hayes' prediction of 34 percent). Altogether, the total contribution of fossil fuels to our energy production would be about 14 percent less than it is today. In terms of CO_2 emissions that is not wildly encouraging—remember the United States is only part of the problem—but it is a step in the right direction.

Unfortunately—in the view of many—this step will not come without a continued donation from nuclear power. Those who argue that nuclear energy should be eliminated completely are simply not being realistic. In some areas of the country—notably in New England, around Chicago, and in parts of the Southeast—nukes generate 35 to 50 percent of all electricity.

If we are to see a future reduction in the role of the fossils, a slow, reasoned expansion of nuclear energy is required. It is not illogical to expect that nuclear could be furnishing about one-tenth of our energy by the end of the century. The Three Mile Island incident forced the nuclear industry to take a good, hard look at itself. As a result, improvements in personnel training and security measures are coming. Future nuclear plants will be constructed away from populated areas (some facilities currently operating near population centers may have to be shut down). And research is continuing on the waste problem. A new process is now being tested that may actually be able to neutralize nuclear wastes.[21]

George J. Church, a senior writer for *Time*, put the issue of the future of nuclear power into perspective in a "Time Essay": "Caution, sobriety, careful weighing of risks, which cannot be escaped, ought to be the watchwords. Slogan shouting—'Hell, no, we won't glow,' vs. 'Let the bastards freeze in the dark'—merely impedes progress toward America's energy future. Simply put, the nation needs to move forward to improve the safety, reliability and efficiency of all forms of energy—including nuclear . . ."[22]

DOING BETTER: NATIONAL GOALS

Assuming that nukes are given a limited part in our energy future, that solar energy can fulfill 20 percent of our needs by 2000, and that our overall energy growth rate slows to less than 1 percent, the energy spectrum at the turn of the century could look like that presented in Table 14—1. I would term the assumptions and numbers in

Table 14—1.   U.S. Energy Sources,[a] 2000.

	Percentage of Total
Oil	26
Gas	14
Coal	30
Nuclear	10
Solar[b]	20

[a] Assumes a 20 percent contribution by solar, and an average annual growth rate of less than 1 percent.

[b] Solar is defined broadly to include all solar-related energy sources, including wind, hydro, ocean, biomass, and so forth.

this scenario a *minimum* goal. With enlightened and brave leadership, we can do much better.

We should set a national target of an actual 20 percent cut in our energy consumption by the end of the century, along with a 25 percent contribution from solar. Such a goal, given the proper incentives and sense of national purpose, is not out of reach. Consider the benefits we would derive from such a program:

1. It would be the quickest and most economical route to energy independence.
2. Our dependence on foreign oil could cease entirely.
3. Coal production could actually be cut to 65 percent of contemporary output.
4. Total fossil fuel consumption could be *halved*!
5. The goal of 25 percent solar could be more easily met than could a 20 percent target at a higher overall consumption rate.
6. Natural gas and nuclear would furnish a greater percentage of our energy, but the absolute energy output needed from each would be no higher than with any other scenario.

Table 14–2 shows what the share of total energy production would be for each source.

It is not an easy set of numbers to achieve. Neither is it an impossible one. It will involve personal commitment to slight changes in our life-styles, particularly as they relate to transportation. We will walk

Table 14–2.   U.S. Energy Sources,[a] 2000.

	Percentage of Total
Oil	24
Gas	21
Coal	14
Nuclear	16
Solar[b]	25

[a] Assumes a 25 percent contribution by solar, and an actual reduction in energy consumption by 20 percent.

[b] Solar is defined broadly to include all solar-related energy sources, including wind, hydro, ocean, biomass, and so forth.

more, carpool more, use mass transit more, and drive more fuel efficient cars. At least until we become more conservation conscious or more dependent upon solar energy, we will keep our homes cooler in the winter and warmer in the summer. We will dress more in keeping with the weather. And we will complain about how much energy is costing us. There is no cheap way. Only less costly ones, and the conservation and solar route offers the lowest rates of all.

We can reinforce our resolve by making certain that our elected officials support programs that boost solar energy and conservation, that diminish our reliance on fossil fuels, that give nuclear power reasonable consideration, and that do not encourage development of synthetic fuels.

We must be active in special interest groups that advocate similar goals. We must back candidates who understand the goals and the reasons behind them, and remove from office (not re-elect) those representatives who cannot see beyond the ends of their elected terms. We have too many such individuals serving us now. After listening to my plea for more long-range political planning on the greenhouse threat and other energy-related environmental issues, a man once asked me, "How in the world can you expect a congressman, whose term of office expires in fifteen months, to be concerned about environmental problems that will crop up twenty years from now?"

But we must expect that. We can not afford to demand any less of our political leaders. We must find forward-thinking, enlightened politicians, encourage them, support them, elect them, and then let them hear our voices. One voice is not loud, but enough of them speaking together can be heard around the world.

Ron Kovic, a Vietnam veteran who returned home paralyzed from his chest down, ended his moving novel *Born on the Fourth of July* with two short sentences. He was reflecting on the personal Arcadia of his own youth: hitting baseballs, having Mickey Mantle for a hero, being in love with young high school girls, and listening to Dell Shannon singing "Runaway" on warm spring days.

At times I find myself wondering whether we might not someday find Ron Kovic's two sentences horribly applicable to our own past environment and energy use: "It was all sort of easy. It had all come and gone."[23]

It does not have to end that way.

By making the right choices now, and by perhaps temporarily surrendering a bit of the economic and social ease we have come to accept as an inalienable right, we can have a better world in which to live by the end of the century. A world with plentiful energy, abundant food, sufficient water, and a clean environment.

Still, no matter what we do, we may not be able to beat the greenhouse threat completely. But we can soften its ultimate effects and delay its onset. We can buy time in which we can learn to adjust to the climatic changes it may foster.

My own hope, though I have no reason now to think that it will be fulfilled, is that fifty years from now the greenhouse threat will be no more than the subject of an old book with yellowing pages, gathering dust on a shelf.

NOTES

Chapter 1

1. "NAS Panel Is Concerned over Atmospheric CO_2 Buildup," *Physics Today* (1977): 17.
2. These figures represent the general range of concentrations established by several studies during the 1970s. See also S. Schneider, and L. Mesirow, *The Genesis Strategy* (New York: Plenum, 1976), p. 179.
3. "NAS Panel Is Concerned."
4. J. Mitchell, "Carbon Dioxide and Future Climate," *EDS* (1977): 3.
5. "NAS Panel Is Concerned," 17.
6. "The Time Is at Hand to View World Affairs Through a Climatic Prism," *Weather and Climate Report* 2 (1979): 3.
7. A. Herzog, *Heat* (New York: Simon and Schuster, 1977).
8. W. L. Gates, *An Essay on the Physical Basis of Climate*, Oregon State University, Climatic Research Institute Report No. 7 (1979), p. 26.
9. P. Shabecoff, "Increase of Carbon Dioxide in Air Alarm Scientists," *New York Times*, 9 June 1979.
10. Ibid.
11. D. Dickson, "U.S. Scientists Warn of Environmental Dangers from Synthetic Fuels," *Nature* 280 (1979): 181.
12. See H. Bernard, *Weather Watch* (New York: Walker and Company, 1979) for a detailed discussion of the 180–year cycle.
13. M. Glantz, "A Political View of CO_2," *Nature* 280 (1979): 189.

Chapter 2

1. These figures represent the general range of concentrations established by several studies during the 1970s. See also S. Schneider, and L. Mesirow, *The Genesis Strategy* (New York: Plenum, 1976), p. 179.

2. The Mauna Loa Observatory measurements are the result of a joint effort between the National Oceanic and Atmospheric Administration (NOAA) and Scripps Institution of Oceanography. Dr. Kirby J. Hanson directs NOAA's Geophysical Monitoring for Climatic Change program; Dr. Keeling—associated with the Mauna Loa program from the outset—is now a professor of oceanography at Scripps.

3. G. Woodwell, "The Carbon Dioxide Question," *Scientific American* 238 (1978): 34.

4. S. Manabe and R.T. Wetherald, "Thermal Equilibrium of the Atmosphere with a Given Distribution of Relative Humidity," *Journal of Atmospheric Sciences* 24 (1967): 241.

5. R. Kerr, "Carbon Dioxide and Climate: Carbon Budget Still Unbalanced," *Science* 197 (1977): 1352. Egon T. Degens, of the Geologisch-Paläontologisches Institut, University of Hamburg, recently noted, however, ". . . improved analytical techniques and instrumental design have added a new dimension to ocean research. Preliminary studies on all fronts suggest that the oceans are the main sink for the man-made CO_2 in our atmosphere. ("Carbon in the Sea," *Nature* 279 (1979): 191.)

6. J. Mason, "Computing Climatic Change," *New Scientist* (1979): 196.

7. J. Mitchell, "Carbon Dioxide and Future Climate," *EDS* (1977): 3; H. Flohn, "Climate and Energy: A Scenario to a 21st Century Problem," *Climatic Change* 1 (1977): 5; R. Rotty, and A. Weinburg, "How Long Is Coal's Future?" *Climatic Change* 1 (1977): 45; and J. Gribbin, "Fossil Fuel: Future Shock?" *New Scientist*, (1978): 541. Some of the most recent work, by Manabe and Ronald J. Stouffer, employing a more sophisticated model of atmospheric circulation and its interaction with the oceans (including sea ice), obtained similar results. Specifically, global warming of roughly 3.7°F was indicated, but with a difference between northern and southern hemispheric warming. Temperature increases of around 4°F were suggested for the northern hemisphere, as against about 3.2°F in the southern hemisphere. ("A CO_2-Climate Sensitivity Study with a Mathematical Model of the Global Climate," *Nature* 282 (1979): 491.)

8. M. Budyko, "Foreign Studies of Contemporary Climate Variations," *Meteorologiya I Gidrologiya* 5 (1978): 112.

9. F. MacIntyre, "On the Temperature Coefficient of P_{CO_2} in Seawater," *Climatic Change* 1 (1978): 349.

10. C. Marchetti, "On Geoengineering and the CO_2 Problem," *Climatic Change* 1 (1977): 59. An example of additional research leading to a bleaker environmental picture was the National Research Council's 1979 conclusion that stratospheric ozone is being depleted at twice the rate computed in a 1976 study. If the ozone layer is substantially diminished by continued widespread use of Freon in spray cans, refrigerators, and plastic foams, more ultraviolet light from the sun will be allowed to reach the earth, causing an increase in the rate of skin cancer.

11. M. Budyko, "Foreign Studies," p. 112.

12. S. Schneider, "On the Carbon Dioxide–Climate Confusion," *Journal of the Atmospheric Sciences* 32 (1975): 2060.

13. Woodwell, "Carbon Dioxide Question."

14. Gribbin, "Fossil Fuel."

15. Rotty and Weinburg, "How Long Is Coal's Future?"

16. R. Bryson and T. Murray, *Climates of Hunger* (Madison: The University of Wisconsin Press, 1977), p. 131; and R. Bryson and G. Dittberner, "A Non-Equilibrium Model of Hemispheric Mean Surface Temperature," *Journal of the Atmospheric Sciences* 33 (1976): 2094.

17. Schneider and Mesirow, *Genesis Strategy*, p. 181; and J. Gribbin, *What's Wrong with Our Weather?* (New York: Charles Scribner's Sons, 1979), p. 151.

18. H. Willet and J. Prohaska, "Patterns, Possible Causes and Predictive Significance of Recent Climatic Trends of the Northern Hemisphere," unpublished paper, The Solar Climatic Research Institute, Inc. 1977.

19. Mason, "Computing Climatic Change."

20. F. Pohl, "Power Play," *Omni*, (1979): 69.

21. J. Hays, "Milankovitch Theory Verified," *Bulletin of the American Meteorological Society* 58 (1977): 184.

22. J. Hays, J. Imbrie, and N. Shackleton, "Variations in the Earth's Orbit: Pacemaker of the Ice Ages," *Science* 194 (1976): 1121.

23. Ibid.

24. Ibid., p. 1121.

Chapter 3

1. H. Lamb, "On the Nature of Certain Climatic Epochs Which Differed from the Modern," *Proceedings of the WMO/UNESCO Rome (1961) Symposium on Climate Changes (Arid Zone XX)*, (New York: UNESCO 1963), p. 125.

2. H. Flohn, *Climate and Weather* (New York: McGraw-Hill, 1969), p. 211.

3. Ibid., p. 210.

4. Lamb, "Climatic Epochs."

5.   H. Bernard, *Weather Watch* (New York: Walker and Company, 1979).

6.   "Geographical Patterns of Climatic Change: 1000BC–1700AD," *Climate Monitor* 6 (1977): 131.

7.   J. Gribbin, "Fossil Fuel: Future Shock?" *New Scientist* (1978): 541.

8.   G. Paltridge, "The Problem with Climate Prediction," *New Scientist* (1979): 194.

9.   Lamb, "Climatic Epochs."

10. H. Willett, "Do Recent Climatic Fluctuations Portend an Immiment Ice Age?" *Geofisica Internacional* 14 (1974): 265.

11. Bernard, *Weather Watch*, p. 20.

12. Willett, "Climatic Fluctuations."

13. L. Hmelevskaja and S. Savina, "Variations of Atmospheric Circulation and Climate in the 20th Century," *Moscow, Akad, Nauk, Priroda* 2 (1969): 38.

14. Ibid.

Chapter 4

1.   R. Felch, "Drought: Characteristics and Assessment," in *North American Droughts*, ed. N. Rosenberg, (Boulder: Westview Press, 1978), p. 25.

2.   H. Bernard, *Weather Watch*, p. 37.

3.   D. Ludlum, *Weather Record Book* (Princeton: Weatherwise, Inc., 1971), p. 39.

4.   J. Cornell, *The Great International Disaster Book* (New York: Pocket Books, 1979), p. 247.

5.   C. Stockton, "Long-Term Spatial and Temporal Drought Frequency Analysis in Western United States Utilizing Tree Rings" *National Science Foundation Report DES74–24163* (1976); and C. Stockton and D. Meko, "A Long-Term History of Drought Occurrence in Western United States as Inferred from Tree Rings," *Weatherwise* 28 (1975): 245.

6.   J. Steinbeck, *The Grapes of Wrath*, (New York: Penguin Books, 1977), p. 2.

7.   Climate & Food (Washington: National Academy of Sciences, 1976), p. 23.

8.   Ibid, p. 109.

9.   Ibid.

10. Ibid.

11. Ibid., p. 127.

12. L. Thompson, "Weather Variability, Climatic Change, and Grain Production," *Science* 188 (1975): 535.

13. "Is U.S. Running Out of Water?" *U.S. News & World Report*, 18 July 1977: 33.

14. Thompson, "Weather Variability."

15. A. Ehrlich and P. Ehrlich, *The End of Affluence* (New York: Ballantine Books, (1974), p. 32. Paul Ehrlich is a Stanford University biologist and author of the bestselling book *Population Bomb. The End of Affluence* is the follow-up to the now classic *Population Bomb.*

16. Cornell, *Disaster Book*, p. 327.

17. "Is U.S. Running Out of Water?"

18. "Water: Time to Start Saving?" *Consumer Reports*, (1978): p. 294.

19. J. Dracup, "Impact on the Colorado River Basin and Southwest Water Supply," in *Climate, Climatic Change and Water Supply* (Washington: National Academy of Sciences, 1977), p. 121.

20. Ibid.

21. Ibid.

22. Ibid.

23. Ibid.

24. R. Gustaitis, "Water!" *Boston Sunday Globe*, 22 January 1978, p. A3.

25. Dracup, "Colorado River Basin," p. 121.

26. Ibid.

Chapter 5

1. D. Ludlum, *Early American Hurricanes 1492–1870* (Boston: The American Meteorological Society, 1963), p. 103.

2. J. McCarthy, *Hurricane* (New York: American Heritage Press, 1969), p. 148; and G. Dunn and B. Miller, *Atlantic Hurricanes* (Baton Rouge: Louisiana State University Press, 1964), p. 274.

3. H. Bernard, "Remember Carol, Diane, Donna, and the L.I. Express?" *The Hartford Courant*, 1 September 1974, p. 2.

4. The lowest pressure ever observed in a tropical cyclone was in Typhoon Tip in the western Pacific Ocean in October 1979. The U.S. Air Force's 54th Weather Reconnaissance Squadron, flying out of Anderson Air Force Base on Guam, measured a central pressure in the "superphoon" of 25.68 inches.

5. Much of the information in this section of the chapter came from the following sources:
Dunn and Miller, *Atlantic Hurricanes;* D. Harris, "Characteristics of the Hurricane Storm Surge" *Technical Paper No. 48* (Washington: Government Printing Office, 1963); D. Ludlum, *Weather Record Book* (Princeton: Weatherwise Inc., 1971); annual hurricane season reviews published in the *Monthly Weather Review* (American Meteorological Society, Boston); and *Weatherwise* (Helen Dwight Reid Educational Foundation, Washington, D.C.).

6. J. Galway, "Some Climatological Aspects of Tornado Outbreaks," *Monthly Weather Review* 105 (1977): 477. Recent (since 1975) statistics were gathered from annual tornado season reviews published in the *Monthly Weather Review*, and *Weatherwise*.

7. Ibid.

8. T. Fujita, D. Ludlum, and A. Pearson, "Long-Term Fluctuation of Tornado Activities," SMRP Paper No. 128 (The University of Chicago, 1975).

9. Information derived from map of tornado tracks, "U.S. Tornadoes 1930–74," University of Chicago, 1976.

10. Statistics in this section gathered from Ludlum, *Weather Record Book*.

Chapter 6

1. H. Willett, "Do Recent Climatic Fluctuations Portend an Imminent Ice Age?" *Geofisica Internacional* 14 (1974): 265.

2. F. Bair and J. Ruffner, Eds, *The Weather Almanac* (Detroit: Gale Research Company, 1977), p. 106.

3. D. Ludlum, *The Weather Record Book* (Princeton, N.J. Weatherwise, Inc., 1971), p. 29.

4. Most of the statistical information in this section came from D. Ludlum, *The Weather Record Book* (Princeton, N.J.: Weatherwise, Inc., 1971), and the April and June 1971 issues of *Weatherwise* (Helen Dwight Reid Educational Foundation, Washington, D.C.).

5. Bair and Ruffner, *Weather Almanac* p. 109.

6. J. Fast, *Weather Language* (New York: Wyden Books, 1979), p. 182.

7. See 4.

8. D. Ludlum, *New England Weather Book* (Boston: Houghton Mifflin, 1976), p. 73.

9. See 4.

10. J. Cornell, *The Great International Disaster Book* (New York: Pocket Books, 1979), p. 167.

Chapter 7

1. G. Schultz, *Ice Age Lost* (Garden City, N.Y.: Anchor/Doubleday, 1974), p. 224.

2. H. Rosendal, "Mexican West Coast Tropical Cyclones, 1947–1961," *Weatherwise* 16 (1963): 226.

3. "Weatherwatch: September and October 1976," *Weatherwise* 29 (1976): 296; and E. Gunther, "Eastern North Pacific Tropical Cyclones of 1976," *Monthly Weather Review* 105 (1977): 508.

Chapter 11

1. H. Bernard, *Weather Watch* (New York: Walker and Company, 1979), p. 97.
2. B. Hayden and D. Resio, "Recent Secular Variations in Mid-Atlantic Winter Extratropical Storm Climate," *Journal of Applied Meteorology* 14 (1975): 1223.
3. P. Jones, P. Kelly, and T. Wigley, "Scenario for a Warm, High-CO_2 World," *Nature* 283 (1980): 17.

Chapter 12

1. H. Lamb, "On the Nature of Certain Climatic Epochs Which Differed From the Modern," *Proceedings of the WMO/UNESCO Rome (1961) Symposium on Climate Changes (Arid Zone XX)*, (New York: UNESCO 1963), p. 125.
2. R. Bryson and T. Murray, *Climates of Hunger* (Madison: The University of Wisconsin Press, 1977), p. 68.
3. Lamb, "Climatic Epochs."
4. S. Schneider and L. Mesirow, *The Genesis Strategy* (New York: Plenum, 1976), p. 69.
5. Lamb, "Climatic Epochs."
6. Schneider and Mesirow, *Genesis Strategy*, and Bryson and Murray, *Climates of Hunger*, p. 42.
7. Schneider and Mesirow, *Genesis Strategy*, p. 70.
8. Lamb, "Climatic Epochs."
9. Ibid.
10. Schneider and Mesirow, *Genesis Strategy*, p. 76; and Bryson and Murray, *Climates of Hunger*, pp. 37, 38, 42.
11. "Geographical Patterns of Climatic Change: 1000BC—1700AD," *Climate Monitor* 6 (1977): 131.
12. J. Mercer, "West Antarctic Ice Sheet and CO_2 Greenhouse Effect: A Threat of Disaster," *Nature* 271 (1978): 321; and K. Rose, T. Sanderson, and R. Thomas, "Effect of Climatic Warming on the West Antarctic Ice Sheet," *Nature* 277 (1979): 355.
13. H. Flohn, *Climate and Weather* (New York: McGraw-Hill, 1969), p. 216.
14. The information in this section was taken from Lamb, "Climatic Epochs," and J. Gribbin, "Fossil Fuel: Future Shock?" and from the map of the Altithermal Period prepared by Dr. William W. Kellogg of the National Center for Atmospheric Research, Boulder, Colorado.
15. D. McLean, "A Terminal Mesozoic 'Greenhouse': Lessons from the Past," *Science* 201 (1978): 401.

16. Rose, Sanderson and Thomas, "Climatic Warming," p. 240.
17. H. Flohn, "Climate and Energy," *Climatic Change* 1 (1977): 17.
18. Ibid.
19. W. Sullivan, "Climatologists Are Warned North Pole Might Melt," *New York Times*, 14 February 1979, p. A21.
20. J. Mason, "Computing Climatic Change," *New Scientist*, 1979, p. 196.
21. G. Schultz, *Ice Age Lost* (Garden City, New York: Anchor/Doubleday, 1974), p. 302.
22. A. Herzog, *Heat* (New York: Simon and Schuster, 1977), p. 248.
23. J. Laurmann, "Market Penetration Characteristics for Energy Production and Atmospheric Carbon Dioxide Growth," *Science* 205 (1979): 896.
24. Schneider and Mesirow, *Genesis Strategy*, p. 100.
25. Conversation with James McQuigg, 22 November 1977.
26. C. Black, "AID Official Warns of Famine in Asia," *Boston Globe*, 22 March 1979.
27. From a seminar given by Dr. Eric Walther, manager of the Food Production Climate Mission, Kettering Foundation, Dayton, Ohio, at Environmental Research & Technology, Inc., Concord, Massachusetts, 5 September 1975.
28. L. McLaughlin, "Feeding the World," *Boston Globe*, 30 April 1977.
29. "The Consequences of a Hypothetical World Climate Scenario Based on an Assumed Global Warming Due to Increased Carbon Dioxide," symposium report, Aspen Institute for Humanistic Studies, 1978, p. 34.
30. Ibid.
31. Schneider and Mesirow, *Genesis Strategy*, p. 288.

Chapter 13

1. Personal correspondence with Lawrence Gates, 9 August 1979.
2. "Can Carbon Dioxide be Removed from the Atmosphere?" IAEA *Bulletin* 21 (1979): 9.
3. J. Botzum and R. Jacobius, eds., "The CO_2 Question Has Provoked Serious Reconsideration of a Vast Synfuels Program," *Weather & Climate Report*, (August 1979): 3.
4. C. Marchetti, "On Geoengineering and the CO_2 Problem," *Climatic Change* 1 (1977): 59.
5. J. Botzum and R. Jacobius, eds., "The CO_2 Question."
6. E. Hayes, "Energy Resources Available to the United States, 1985 to 2000," *Science* 203 (1979): 233.
7. A. Parisi, "Fuel Savings Are Growing But Not Very Energetically," *New York Times*, 4 March 1979.

8. An article entitled "Chinese Scientists Envisage Coal as Key Energy Source of Future" in the January 13, 1980, *New York Times* with a dateline of Peking stated, "Scientists at a national energy conference here have agreed that coal will be China's major energy source in the future, the New China News Agency said today."

9. Hayes, "Energy Resources."

10. Ibid.

11. Project Interdependence, *U.S. and World Energy Outlook Through 1990* (Government Printing Office, Washington, D.C. 1977).

12. "U.S. Energy Aide Predicts 'Misery' for Oil Consumers," *Boston Globe*, 31 January 1979.

13. M. Shanahan, "Panel says Saudi Oil Will Fall Short," *Boston Globe*, 15 April 1979.

14. Hayes, "Energy Resources," p. 233.

15. J. Williams, ed. *Carbon Dioxide, Climate and Society*, (New York: Pergamon Press, 1978), p. 264.

16. "The Limits of Coal," *Boston Globe*, 5 July 1979, p. 10.

17. Ibid.

18. T. O'Toole, "Premature Deaths Laid to Pollutants," *Boston Globe*, 5 July 1977, p. 36.

19. B. Mohl, "Coal—the Alternative Stymied by Regulations," *Boston Globe*, 16 April 1979, p. 1.

20. G. Church, "Looking Anew at the Nuclear Future," *Time* (1979): 32.

21. H. Bethe, "The Necessity of Fission Power," *Scientific American* 234 (1976): 21.

22. "Ultimate Solution," *Parade*, 2 September 1979, p. 8.

23. L. Carter, "Policy Review Boosts Solar as a Near-Term Energy Option," *Science* 203 (1979): 252.

24. Barry Commoner on NBC's *Today Show*, 22 June 1979.

25. "Energy Policy: 'How Do I "Unclap?"'" *Government Executive* (1979): 32.

26. Carter, "Policy Review"; and "'Solar Society' Does Not Need 'Moral Equivalent of War,'" *Solar Energy Intelligence Report*, 29 January 1979, p. 42.

27. "Solar Advocates Allies of Oil Execs; U.S. 20% Solar by Year 2000: Harvard," *Solar Energy Intelligence Report*, January 29, 1979, p. 43.

28. "Solar Sell," *Time*, 1979.

29. R. Levy, "Search Is Leading Us into Past—Our Rivers," *Boston Globe*, 17 April 1979.

30. B. Mohl, "To Some, Wind Power Is the Ultimate Answer," *Boston Globe*, 21 April 1979.

31. M. Gustavson, "Limits to Wind Power Utilization," *Science* 204 (1979): 13.

32. J. Reed, "Wind Power Climatology," *Weatherwise* 27 (1974): 237.

33. S. Turner, "Star in the Sky," *New England*, (1977): 20.

34. Ibid.

35. "Energy: Fuels of the Future," *Time* (1979): 72.

36. R. Cooke, "Geothermal: The Energy Buried in the Earth," *Boston Globe*, 20 April 1979, p. 6.

37. R. Cooke, "Jules Verne's Answer was Hydrogen," *Boston Globe*, 22 April 1979.

38. "Energy Consumption per Unit of GNP," *The Hammond Almanac* (Maplewood, N.J.: Hammond Almanac, Inc., 1978), p. 208.

39. Warren Johnson on NBC's *Today Show*, 30 May 1979.

40. The figures for 1980 and 1985 refer to fleet averages for cars manufactured in those years. The nationwide "real" average will not reach those levels until all older, less efficient cars are off the road.

41. P. Abelson, "Energy Conservation," *Science* 204 (1979).

42. D. Yergin, "The Economics of Conservation as a Source of Energy," *Boston Globe*, 23 April 1979, p. 10.

43. P. Abelson, "Energy Conservation."

Chapter 14

1. L. Carter, "A Warning on Synfuels, CO_2, and the Weather," *Science* 205 (1979): 376.

2. "The CO_2 Question Has Provoked Serious Reconsideration of a Vast Synfuels Program," *Weather & Climate Report*, (August 1979): 3.

3. J.M. Mitchell, "Carbon Dioxide and Future Climate," *EDS* (March 1977): 3.

4. Personal correspondence with Kirby Hanson, 13 August 1979.

5. "The CO_2 Problem Should Not Be the Basis for a Decision Against Synfuels," *Weather & Climate Report*, 2 (September 1979): 1.

6. D. Yergin, "It's Now or Never for Conservation," *Boston Globe*, 13 November 1979, p. 19.

7. R. Stobaugh and D. Yergin, eds. *Energy Future* (New York: Random House, 1979), p. 176.

8. Such programs would offer tax rebates, tax credits, and low cost loans to conservation-oriented consumers; continue to mandate strong automobile fuel efficiency standards; and urge revision of building codes and mortgage loan requirements in such a manner as to facilitate energy-efficient construction.

9. P. Hager, "Portland—a City Takes on the Energy Crisis," *Boston Globe*, 13 October 1979.

10. L. Cannon and J. Kotkin, "The Growth of the West," *Boston Globe*, 1 August 1979, p. 2.

11. N. Pierce, "Going for a Free Ride in Seattle," *Boston Globe*, 29 October 1979.

12. "The Mess in Mass Transit," *Time*, (1979): 52.

13. Stobaugh and Yergin, *Energy Future*, p. 147.

14. Ibid., p. 212.

15. Cannon and Kotkin, "Growth of the West."

16. Duane Day speaking at an energy seminar in Stow, Massachusetts, 12 September 1979.

17. Stobaugh and Yergin, *Energy Future*, p. 213.

18. L. Carter, "Policy Review Boosts Solar as a Near-Term Energy Option," *Science* 203 (1979): 252; and " 'Solar Society' Does Not Need 'Moral Equivalent of War,' " *Solar Energy Intelligence Report*, 29 January 1979, p. 42.

19. Personal conversation with Dr. J.S. McNiel, Jr., President, Mobil Tyco Solar Energy Corporation, 15 November 1979.

20. E. Hayes, "Energy Resources Available to the United States, 1985 to 2000," *Science* 203 (1979): 233.

21. Personal conversation with David Rockford, nuclear security coordinator at Stone and Webster's Cherry Hills, New Jersey Operations Center, 8 November 1979.

22. G. Church, "Looking Anew at the Nuclear Future," *Time*, (1979).

23. R. Kovic, *Born on the Fourth of July* (New York: McGraw-Hill, 1976), p. 208.

INDEX

Albuquerque, New Mexico
  climate of 1930s compared to
    current climate, 84–85, 86
Altithermal Period
  as climatic analog for greenhouse
    effect, 131–133
  as climatic analog in Africa, 139
  as climatic analog in India, 139
  as climatic analog in Russia, 139
  as climatic analog in United States,
    139
  in Arctic, 132
  in Europe, 132–133
  in North America, 132–133
  in Russia, 132–133
  in Southern Hemisphere, 132, 133
Andrus, Cecil D.
  on movement of water supplies from
    one state to another, 50
Anthropogenic dust
  and climate change, 22
Anthropogenic heat
  effect of direct input of on climate, 23
Apalachicola, Florida
  climate of 1930s compared to
    current climate, 116, 118
Arctic Ocean
  implications of ice-free, 134–135

Bagge, Carl
  on federal regulations governing
    coal industry, 148
Baton Rouge, Louisiana
  climate of 1930s compared to
    current climate, 106–108
Biomass energy, 152
Bismarck, North Dakota
  climate of 1930s compared to
    current climate, 87, 89–90
Boston, Massachusetts
  climate of 1930s compared to
    current climate, 109–111, 112
British Meteorological Office
  climate model assuming ice-free
    Arctic Ocean, 135
Broecker, Wallace
  on "super-interglacial," 7
Bryson, Reid
  atmospheric dust—climate model,
    22
  on drought in the Midwest, 130
Bureau of Reclamation
  cloud seeding, 50
Burlington, Vermont
  climate of 1930s compared to
    current climate, 109–110

California
  solar energy applications in,
    164-165
  tax credits in for solar energy,
    164-165
Carbon dioxide
  atmospheric, 1
  climatic warming induced by, 6-8
  effect of doubling concentration
    in atmosphere, 18
  exponential nature of increase of
    in atmosphere, 14-16
  increase of in atmosphere, 1-2
  land biota as net source of, 16
  measurements of in atmosphere,
    14-15
  physical properties of, 6
  removing excess from atmosphere,
    141-143
  systematic oscillation in annual
    concentrations of in atmosphere,
    15-16
Carter, President Jimmy
  solar energy goals, 151
  synthetic fuels program, 8
Central Arizona Project, 47, 50
Church, George J.
  on future of nuclear energy, 167
Cincinnati, Ohio
  climate of 1930s compared to
    current climate, 101-103, 106
CLIMAP, 24
Climate models
  and CO_2 problem, 17-18
Climatic analogs, 31-32, 33-34
Climatic feedback, 19
Climatic Research Unit
  climatic analog for future high-CO_2
    world, 124-126
Climatic stress, 32, 57
Cloud seeding, 50
Clusen, Ruth
  on CO_2 and synthetic fuels, 160
Coal
  as energy source in China, 144
  future domestic production of, 166
  problems associated with use of,
    147-149
  U.S. reserves of, 147
Cold waves
  during 1930s, 64-67

Columbia, Missouri
  climate of 1930s compared to
    current climate, 94-96
Commoner, Barry
  on President Carter's solar program,
    151
Corn
  effects of drought on, 41-42
Council on Environmental Quality
  goal of, 151
Cycles, climatic
  masking effects of on anthropogenic
    warming, 9-11
  180-year cooling, 8-9

Dawson, John M.
  on future of nuclear fusion, 150
Day, Duane
  on tax credits for solar energy
    applications, 165
Denver, Colorado
  climate of 1930s compared to
    current climate, 82-83, 84
  mass transit in, 164
Department of Energy
  cost goals for photovoltaic produc-
    tion, 151
Department of Interior
  energy development and water con-
    sumption, 48
Desalination of sea water, 50
Detroit, Michigan
  climate of 1930s compared to
    current climate, 101-104, 108
Dittberner, Gerald
  atmospheric dust—climate model,
    22
Dodge City, Kansas
  climate of 1930s compared to
    current climate, 94-95
Dow Chemical
  energy conservation by, 157
Drought
  and forest fires, 45-46
  cycle of in western U.S., 35-36
  definition of, 35
  during Medieval Warm Period,
    129-130
  effect of on Colorado Basin water
    supplies, 46-49
  in Australia, 33

in Midwest, 33
in Russia, 33
in 1950s, 39
in 1970s, 39-40
resulting from ice-free Arctic Ocean,
   134
Dust bowl, 37-41, 43

East Antarctic Ice Sheet
melting of, 130-131
Ehrlich, Anne and Paul
The End of Affluence, 45
Electricity
costs of producing, 148-149
Energy
projected U.S. sources of in 2000,
   144
U.S. sources of as of 1977, 143
Energy conservation, 156-157
as an answer to the energy crisis,
   162-163
Energy Policy and Conservation Act
of 1975, 156
Energy programs
and elected officials, 169
national goals, 167-169
Eric the Red, 128

Fairbanks, Alaska
climate of 1930s compared to
   current climate, 77-78
Flohn, Herman
on potential water supply problems
   in California and Utah, 134
Floods
during 1930s, 67-69
Food supplies
and necessity of redefining
   agricultural areas, 139-140
future of, 138-140
threatened by greenhouse effect,
   134
Fujita, T. Theodore
on tornado death rate anomalies, 55

Galveston, Texas
climate of 1930s compared to
   current climate, 96-99
Gates, W. Lawrence
on potential climatic
   catastrophe, 7
Geothermal energy, 154-155

Gillette Company
energy conservation by, 157
Glantz, Michael
on CO_2 problem, 12
Great New England Hurricane, 52
Greenhouse effect, 6, 16
and melting ice caps, 7, 8
hemispheric perspective by Climatic
   Research Unit, 124-126
in an amalgam of issues, 161-162
Greensboro, North Carolina
climate of 1930s compared to
   current climate, 114-115
Gustavson, M.R.
on wind power, 153

Hanson, Kirby J.
on scientists as policy makers, 160
Havre, Montana
climate of 1930s compared to
   current climate, 87, 89-90
Hayes, Earl T.
on future of solar energy, 151
on future U.S. production of
   natural gas, 146
on slowing of U.S. energy growth
   rate, 163
prediction of U.S. annual energy
   growth rate, 143
Hays, John D.
on ice ages and changes in earth's
   orbital geometry, 24-25
"Heat island" effect, 72-73
Heat waves
dangers of, 63-64
during 1930s, 59-63
pollution build-up during, 63
Henault, Peter
on energy conservation, 164
Heronemus, William E.
on wind power, 153, 154
Herzog, Arthur
Heat, 135
Hurricane Camille, 54
Hurricanes
near U.S. East and Gulf coasts
   during 1930s, 51-54
near U.S. West coast during 1930s,
   76
Hydroelectricity, 152-153
Hydrogen-derived energy, 155-156

Ice Age, 23-24

Jet stream
  shift of as earth's climate changes,
    27-31
Johnson, Warren
  on energy conservation, 156

Keeling, Charles David
  and Mauna Loa program, 13-14
Kellogg, William W.
  on using real earth for climate
    model, 31
Kelly, John J., Jr.
  making Arctic CO_2 measurements,
    13
Kovic, Ron
  Born on the Fourth of July, 169

Labor Day Storm of 1935, 54
Lakeland, Florida
  climate of 1930s compared to
    current climate, 116, 119
Lamb, Hubert H.
  on upper winds associated with
    Little Ice Age and Medieval
    Warm Period, 31-32
Lincoln, Nebraska
  climate of 1930s compared to
    current climate, 87-90, 93
Little Ice Age, 28, 31
Los Angeles, California
  climate of 1930s compared to
    current climate, 73, 74-76
Ludlum, David
  on New England floods of 1936, 68

MacDonald, Gordon J.
  "The Long-term Impact of
    Atmospheric Carbon Dioxide
    on Climate" (report), 7
Maidique, Modesto A.
  on tax credits for solar energy
    applications, 165
Manabe, Syukuro
  on global temperature increase due
    to atmospheric CO_2 buildup, 16
Marchetti, Cesare
  on climatic feedbacks, 19
Market penetration time
  as applied to energy sources, 137

Mason, John
  on climate models, 18
Mass transit
  future of, 164
Mauna Loa, 13-14
McQuigg, James
  on difference between good and
    bad harvest year, 138
Medieval Warm Period, 28, 31
  as climatic analog for greenhouse
    effect, 130
  in Europe, 127-129
  in Greenland, 128
  in Iceland, 128
  in U.S., 129-130
Memphis, Tennessee
  climate of 1930s compared to
    current climate, 106-107
Mesozoic Era, 133
Midland-Odessa, Texas
  climate of 1930s compared to
    current climate, 96-98
Milankovitch, Milutin
  theory relating climatic change to
    earth's orbital behavior, 23-25
Milton (Boston), Massachusetts
  climate of 1930s compared to
    current climate, 109-111, 112
Minneapolis-St. Paul, Minnesota
  climate of 1930s compared to
    current climate, 87, 89-91
Mitchell, J. Murray, Jr.
  on CO_2 problem, 160

National Academy of Sciences
  climatic change and agricultural
    pests, 43
  future atmospheric CO_2 concentra-
    tions, 6
  impact of changing weather patterns,
    7
  water supply augmentation, 50
National Severe Storms Forecast
    Center, 54
Natural gas
  future domestic production of, 166
  U.S. production of, 146
Natural Gas Policy Act, 146
New York, New York
  climate of 1930s compared to
    current climate, 109-110, 113

1930s
as climatic analog for CO_2 -induced warming, 33–34
climatic change during, 32
climatological conditions of as representative of initial stages of greenhouse effect, 69–70, 71, 118
summary of climatological conditions during, 119–123
wind patterns associated with climatological conditions during, 123–124
Nuclear energy
assessment of risks associated with, 150
future of, 149–150, 166–167
nuclear fusion, 150
opposition to, 149
serious questions about, 149–150

Ocean-derived energy, 154
Oceans
warming of during 1930s, 33
Ogallala Aquifer
depletion of, 44
Oil
future domestic production of, 166
future shortages of, 145
reserves of in U.S., 144–145
U.S. dependence on foreign sources of, 145–146
world production of, 145
Oklahoma City, Oklahoma
climate of 1930s compared to current climate, 94–95
O'Leary, John
on financial burden of oil on world economy, 145
OPEC, 145
Oregon
tax credits in for alternative energy sources, 165
Osburn, James E.
on future decline in irrigated farming, 44

Paltridge, Garth
on correlating expected climatic trends with past behavior, 31
Phoenix, Arizona
climate of 1930s as compared to current climate, 84–86

Photovoltaic production
cost of, 151
Pohl, Frederik
on thermal pollution of atmosphere, 23
Polar regions
amplification of climatic warming in, 18
Portland General Electric, 165
Portland, Oregon
energy conservation program of, 163–164
mass transit in, 164
Press, Frank
on CO_2 problem, 159
Project Interdependence, 145

Reed, C.D.
on July 1936 heat in Iowa, 60
Revelle, Roger
on CO_2 -induced rise in atmospheric temperature, 159
on potential use of land for agriculture, 138
Rochester, New York
climate of 1930s compared to current climate, 109–111
Rockford, Illinois
climate of 1930s compared to current climate, 101–103, 105
Rome, Georgia
climate of 1930s compared to current climate, 116–117
Rotty, Ralph
on control measure for CO_2, 142
on scrubbing CO_2 from waste gases, 142

Salem, Oregon
climate of 1930s compared to current climate, 73–74, 75
Salt Lake City, Utah
climate of 1930s compared to current climate, 79–81
San Francisco, California
climate of 1930s compared to current climate, 73, 74–75
Sault Sainte Marie, Michigan
climate of 1930s compared to current climate, 101–103

Schneider, Stephen H.
  on long-term food supply outlook,
    140
  on planning future fossil fuel con-
    sumption, 20
Sea level
  rise of threatened by greenhouse
    effect, 134, 135-136
Seattle, Washington
  energy conservation drive in, 164
  mass transit passes issued in, 164
Sheridan, Wyoming
  climate of 1930s compared to
    current climate, 82-83
Slade, David H.
  on potential for serious problem re.
    atmospheric CO_2, 7-8
Snowfall
  recent increase in around eastern
    Great Lakes, 111-113
Solar energy, 150-151
  federal financial support required
    for, 165-166
Soybeans
  effects of drought on, 45
Steinbeck, John
  The Grapes of Wrath, 40-41
Stoler, Peter
  on technical writing, 8
Sulfur dioxide
  EPA regulations regarding, 148
  scrubbing of from waste gases, 142
Sullivan, John
  on future shortfall of food supplies,
    138
Synthetic fuels
  cost of compared to conservation,
    161
  "dirtier" in terms of CO_2, 8
  effect of on water supplies, 48-49
  environmental problems associated
    with, 48

Tax credits
  as incentive for solar energy,
    164-165
Technology
  as an answer to the energy crisis,
    162
Temperature, global
  projected trend in, 20-21

Thermometer exposure
  importance of consistency in for
    climatological purposes, 73
Thompson, Louis M.
  on crop yield reductions due to
    weather effects, 44
Three Mile Island, 149, 167
Tillamook Burn, 45
Tornadoes
  during 1930s, 54-56
Tropical Storm Kathleen, 77

Valentine, Nebraska
  climate of 1930s compared to
    current climate, 87, 89-90, 92

Waco, Texas
  climate of 1930s compared to
    current climate, 96-98
Walla Walla, Washington
  climate of 1930s compared to
    current climate, 73, 79-80
Washington, D.C.
  climate of 1930s compared to
    current climate, 114-115
Water Resources Council
  future water shortages, 46
Water supplies
  threatened by greenhouse
    effect, 137
West Antarctic Ice Sheet
  melting of, 130-131
Westerly wind circulation, 27-31
Wetherald, Richard
  on global temperature increase due
    to atmospheric CO_2 buildup, 16
Wheat
  effect of drought on, 43-44
White, Robert M.
  on effects of climate change, 7
Willett, Hurd
  on climatic stress, 57
  on recent hemispheric cooling, 22
  on westerlies during 1930s, 32
Wind energy, 153-154
Winnemucca, Nevada
  climate of 1930s compared to
    current climate, 80, 83

Yergin, Daniel
  on energy conservation, 156, 163

ABOUT THE AUTHOR

Mr. Bernard is a consulting meteorologist and author. He is acting president of the Greater Boston Chapter of the American Meteorological Society (AMS). He is a professional member of the AMS, The National Weather Association, The Authors Guild, and The Word Guild. He is the author of *Weather Watch*, a book for the layman on climate change and climate trends in the U.S. over the next several decades. He has also written a number of articles on weather and climate which have been published in both regional and national publications. During 1978–79 he was manager of a round-the-clock, worldwide forecasting operation at Environmental Research & Technology, Inc. (ERT), an environmental consulting firm located in Concord, Massachusetts. At ERT Mr. Bernard also served as a senior staff meteorologist in applied meteorology, and as deputy supervisor of the data analysis and reports department. Prior to joining ERT Mr. Bernard was a staff meteorologist with the Travelers Weather Service in Hartford, Connecticut, and a weather officer in the United States Air Force.